STUDIES IN PROBABILITY THEORY

MAA STUDIES IN MATHEMATICS

Published by
THE MATHEMATICAL ASSOCIATION OF AMERICA

Studies in Mathematics

The Mathematical Association of America

Volume 1: STUDIES IN MODERN ANALYSIS
edited by R. C. Buck

Volume 2: STUDIES IN MODERN ALGEBRA
edited by A. A. Albert

Volume 3: STUDIES IN REAL AND COMPLEX ANALYSIS
edited by I. I. Hirschman, Jr.

Volume 4: STUDIES IN GLOBAL GEOMETRY AND ANALYSIS
edited by S. S. Chern

Volume 5: STUDIES IN MODERN TOPOLOGY
edited by P. J. Hilton

Volume 6: STUDIES IN NUMBER THEORY
edited by W. J. LeVeque

Volume 7: STUDIES IN APPLIED MATHEMATICS
edited by A. H. Taub

Volume 8: STUDIES IN MODEL THEORY
edited by M. D. Morley

Volume 9: STUDIES IN ALGEBRAIC LOGIC
edited by Aubert Daigneault

Volume 10: STUDIES IN OPTIMIZATION
edited by G. B. Dantzig and B. C. Eaves

Volume 11: STUDIES IN GRAPH THEORY, PART I
edited by D. R. Fulkerson

Volume 12: STUDIES IN GRAPH THEORY, PART II
edited by D. R. Fulkerson

Volume 13: STUDIES IN HARMONIC ANALYSIS
edited by J. M. Ash

Volume 14: STUDIES IN ORDINARY DIFFERENTIAL EQUATIONS
edited by Jack Hale

Volume 15: STUDIES IN MATHEMATICAL BIOLOGY, PART I
edited by S. A. Levin

Volume 16: STUDIES IN MATHEMATICAL BIOLOGY, PART II
edited by S. A. Levin

Volume 17: STUDIES IN COMBINATORICS
edited by G.-C. Rota

Volume 18: STUDIES IN PROBABILITY THEORY
edited by Murray Rosenblatt

Volume 19: STUDIES IN STATISTICS
edited by R. V. Hogg

M. Kac
Rockefeller University

J. Kiefer
Cornell University

M. R. Leadbetter
University of North Carolina, Chapel Hill

Donald S. Ornstein
Stanford University

G. C. Papanicolaou
Courant Institute, New York University

Murray Rosenblatt
University of California, San Diego

Studies in Mathematics

Volume 18

STUDIES IN PROBABILITY THEORY

Murray Rosenblatt, editor
University of California, San Diego

Published and distributed by
The Mathematical Association of America

Library of Congress Catalog Card Number 78-71935

Complete Set ISBN 0-88385-100-8
Vol. 18 ISBN 0-88385-118-0

Printed in the United States of America

Current printing (last digit):

10 9 8 7 6 5 4 3 2 1

INTRODUCTION

It is hoped that the six articles in this volume give some idea of the range of ideas and current interests in probability theory. This will be only a sparse sampling of the many directions in which probability theory has developed recently. There has always been a strong interplay with applications and some of the papers show this.

In the first paper J. Kiefer deals with some problems of statistical inference in which sequential methods are used. The basic models in mathematical statistics are probability models. In the usual context, the number of observations or "sample size" is fixed, but there are a number of statistical "designs" in which there was found to be some advantage in having a random sample size. Kiefer's article deals with classical sequential methods. More recent work is discussed in the book of Chow, Robbins, and Siegmund: *Great Expectations: The Theory of Optimal Stopping.*

The second paper is written by myself and concerns itself with notions of ergodicity and mixing for stationary processes. Stationary processes are random sequences whose probability structure is invariant under time shifts. The concepts of ergodicity and mixing originally were motivated by questions in statistical mechanics concerning the interchangeability of "time" and "ensemble" averages. The original concepts of ergodicity and mixing are strengthened and related to measures of dependence and asymptotic independence for random sequences. Such measures turn out to be of some use in getting conditions for the validity of the central limit theorem (asymptotic normality for averages) for random processes.

The paper of M. R. Leadbetter first surveys the classical results concerning the asymptotic distribution of maxima of independent sequences. He then considers in some detail conditions under which the same types of results are valid for stationary dependent sequences. These conditions involve appropriate notions of weak mixing.

The paper of G. C. Papanicolaou deals with the asymptotic analysis of differential equations with stochastic (random) coefficients. Here one finds a familiar situation in which probabilistic questions are posed and tools from a number of areas are used, in this case, for example, from operator theory and the theory of differential equations. Problems of the type dealt with in this article arise in a number of areas, one of them being the problem of scattering (of light or radio waves, for example) through an inhomogeneous or random medium.

M. Kac returns to the source of many probabilistic questions in his paper. Some models are constructed so as to give some insight into phase transitions. The paper is a translation of the article "Quelques problèmes mathématiques en physique statistique."

In the last article D. Ornstein is concerned with the isomorphism problem. Recent work of Kolmogorov and Sinai making use of a concept of entropy resolved a number of open questions concerning the equivalence in a certain sense of random stationary sequences. This was the impetus in opening a new area in which Ornstein himself has been a prime contributor.

In reading through some of the articles, it may be helpful to go through the article at first in a more cursory manner so as to pick up a rough feel of the basic ideas and concepts and then to read through the exposition again in a detailed manner. The following list of texts is given as a source that could be useful to a reader:

L. Breiman, *Probability*, Addison-Wesley, Reading, Mass., 1968.

W. Feller, *An Introduction to Probability Theory and Its Applications*, vols. 1 and 2 (3rd ed. v. 1, 1968, 2nd ed. v. 2, 1971), John Wiley, New York.

S. Karlin and H. M. Taylor, *A First Course in Stochastic Processes*, Academic Press, New York, 1975.

M. Rosenblatt, *Random Processes*, Springer-Verlag, New York, 1974.

There are many other directions in probability theory that are not mentioned in the papers of this volume. One particular area that has drawn much attention in recent years is that of potential theory. Some idea of the basic concepts developed in probabilistic potential theory can be gathered by looking at the little volume of E. B. Dynkin and A. A. Yushkevich, *Markov Processes, Theory and Problems*, Plenum Press, New York, 1969.

MURRAY ROSENBLATT

CONTENTS

SEQUENTIAL STATISTICAL METHODS

*J. Kiefer**

1. INTRODUCTION

Most of the standard techniques of statistical analysis are based on experiments in which the amount of experimentation to be conducted is decided before the experiment begins. For example, in trying to select the best of several drugs to cure a disease, the medical research worker might decide in advance how many patients will be treated with each drug. An alternative occurs naturally to the experimenter who finds the resulting data inconclusive: perhaps he should try the drugs on some additional patients. On the other hand, if he does not have the facilities to treat all the patients at once and finds conclusive enough evidence to select the best drugs after observing only a fraction of the intended total number, perhaps he should save time and money by terminating the experiment early. Thus, he decides to lengthen or shorten the experiment, based on the

* Supported by NSF Grant MCS 75-22481.

outcome of the part of the experiment conducted up to the moment of decision. Such a technique is called a *sequential procedure.*

With such natural impetus, experimenters undoubtedly conducted sequential experiments long before the establishment of probability theory as a mathematical tool with which precise properties of experiments could be studied. Even to this day, whether out of ignorance, sloppiness, or deception, some experimenters publish results obtained from sequential experiments, as though the experiments were not of that nature. As we shall see, the meaning of a given set of results on two drugs, each applied to ten patients, is quite different, if the sample size of twenty was chosen in advance and adhered to, from what it is if the experimenter had not specified his sample size in advance but simply continued treating patients until the results looked conclusive, which turned out to occur after twenty patients had been observed. Such subtleties, as well as the definition and calculation of meaningful characteristics of sequential procedures, require the calculus of probability and in particular the study of random walks. This article illustrates the way in which various parts of probability theory arise as the tools for studying sequential procedures.

The formal mathematical study of sequential procedures was initiated during World War II by the Statistical Research Group at Columbia University, under the leadership of Abraham Wald. His work *Sequential Analysis* (1947) has served as the source book for many practical applications of the techniques and as the starting point for a great deal of additional theoretical development in the succeeding thirty years.

The motivation for developing sequential techniques during the war was that those techniques saved considerably over the classical nonsequential methods in the number of observations needed in sampling inspection of various mass-produced items in order to achieve specified standards of quality. This idea of saving observations will be illustrated in a common statistical setting in Section 3, as will be a second motivation: there are some statistical problems that cannot be handled at all, let alone efficiently, by nonsequential methods. For now, we illustrate both of these notions in an example chosen for its elementary probability structure rather than for practicality.

Suppose the value of an integer-valued physical parameter is to be guessed at from independent measurements $X_1, X_2, \ldots,$ each of which takes on the value $\theta - 1$ or $\theta + 1$ with probability 1/2 each if the actual parameter value is θ. This true value of the parameter is unknown to the experimenter. If N observations $X_1, X_2, \ldots, X_N$ are taken, an integer-valued function $t_N(X_1, \ldots, X_N)$ is used to estimate the true value θ of the parameter, and we suppose the experimenter views the outcome as satisfactory only if t_N takes on exactly the value θ. By chance the X_i may turn out such that $t_N = \theta$ or $t_N \neq \theta$. In order to make clear our computation of the chance that an estimator performs unsatisfactorily, we must adopt a notation that makes precise *which* of the infinite family of probability mechanisms (corresponding to different values of the true θ) actually governs the X_i in terms of which the event under consideration is written. In statistical models, where a class of probability laws is under consideration, we denote by $P_\theta\{A\}$ the probability of an event A when θ is the true parameter value, using E_θ for the corresponding expectation operation. Then $P_\theta\{t_N(X_1, X_2, \ldots, X_N) \neq \theta\}$ is a measure of the inaccuracy of the procedure t_N in the present example, often called the *risk function* of the procedure, as a function of θ.

Suppose B_n is the experiment which, for a positive integer n chosen in advance of observing the X_i, takes exactly n measurements $X_1, \ldots, X_n$. This is the *nonsequential* or *fixed sample size* experiment with sample size $N = n$. Define $U_n = \min(X_1, \ldots, X_n)$ and $V_n = \max(X_1, \ldots, X_n)$. The estimator $t'_n(X_1, \ldots, X_n) = U_n + 1$ can be shown to yield a maximum (over θ) risk that is as small as that of any other estimator, and for it we have

$$P_\theta\{t'_n \neq \theta\} = P_\theta\{X_1 = X_2 = \cdots = X_n = \theta + 1\} = 2^{-n}. \tag{1.1}$$

This can be made as small a positive value as the experimenter specifies, by taking n sufficiently large; but of course he must pay the expense of taking n observations. If the cost of experimentation is proportional to the number of observations taken (often but not always the case), the cost of this experiment is then cn where c is the cost per observation.

For a sequential procedure, the number of observations N is a random variable. (Here and in the sequel, the fixed sample size procedure based on n observations is viewed as a degenerate member of the class of all sequential procedures, with $P_\theta\{N = n\} = 1$ for all θ.) If each observation again costs c, the *expected* cost is $cE_\theta N$ when θ is true. Let N^* be the first n for which $V_n - U_n = 2$, and let B^* be the procedure which stops after N^* observations and estimates θ by $t'_{N^*} = U_{N^*} + 1$. Since $V_{N^*} - U_{N^*} = 2$, we obtain $P_\theta\{t'_{N^*} \neq \theta\} = 0$. Furthermore, for $m \geqslant 2$ we have

$$P_\theta\{N^* = m\} = P_\theta\{X_1 = X_2 = \cdots = X_{m-1} \neq X_m\} = 2^{1-m},$$

so that

$$E_\theta N^* = \sum_{m=2}^{\infty} m 2^{1-m} = 3. \tag{1.2}$$

(This series may be summed as

$$\left[\frac{d}{dz}\sum_2^\infty z^m\right]\Bigg|_{z=1/2} = \left[\frac{d}{dz} z^2(1-z)^{-1}\right]\Bigg|_{z=1/2}.)$$

Thus, if $n > 3$, the procedure B_n is inferior to B^* in both respects, that the former results in a higher probability of yielding an incorrect estimate of θ, and also entails on the average taking more observations. In addition, we see that, if we required a procedure for which the probability of an incorrect estimate is 0, for no fixed sample size n would B_n suffice, but B^* would.

This example, extreme for the sake of simplicity, nevertheless illustrates a phenomenon that often enables sequential procedures to be superior to nonsequential ones: there can be "lucky observations." If $V_2 - U_2 = 2$, the procedure B^* takes advantage of the resulting perfect knowledge of θ to stop earlier than B_n does for $n \geqslant 3$; and if $U_n = V_n$, an unlucky circumstance in which B_n stops, B^* goes on to collect more information. In more realistic examples, the possible states of knowledge will not be so extreme and transparent, but the idea of lucky observation sequences remains.

Incidentally, the example also illustrates that the much used estimator $t''_n = n^{-1}\sum_1^n X_i$, the "sample mean," is not appropriate for all models; it performs much worse than t'_n in the above setting.

Next, to illustrate the way in which the interpretation of an outcome depends on the sampling rule used to achieve it, we consider another model with independent, identically distributed random variables (i.i.d. r.v.'s) $X_1, X_2, \ldots$, this time each X_i being "uniformly distributed" from 0 to an unknown parameter value $\theta > 0$, so that each X_i has continuous distribution function (df) given by

$$P_\theta\{X_i \leqslant x\} = \begin{cases} 0 & \text{if } x < 0, \\ x/\theta & \text{if } 0 \leqslant x \leqslant \theta, \\ 1 & \text{if } x > \theta; \end{cases} \tag{1.3}$$

the "uniform" density refers to the constant value θ^{-1} of (d/dx) $P_\theta\{X_i \leqslant x\}$ on $0 < x < \theta$. Thus, θ is an unknown upper bound on the possible value of an observable X_i, and the statistician wants to make some inference about θ. A type of inference somewhat different from estimation of a parameter as considered earlier is *hypothesis testing*, exemplified here as the process of guessing, when it is known only that $\theta \geqslant 1$, whether the true θ is 1 or is >1 (these being the two hypotheses). A statistical procedure t_n based on $X_1, \ldots, X_n$ now makes one of two assertions g_0 and g_1, where these denote, respectively, the guesses that "the true θ is 1" and that "the true θ is >1." By chance the X_i's can yield a correct or incorrect guess for a given procedure (just as in the previous example), and we see that there are two types of incorrect inference that are possible, guessing (from observations on the X_i's) g_0 when the true θ is >1, or guessing g_1 when $\theta = 1$. We would like both types of error to be unlikely; but for any specified number of observations, it is easy to see that a procedure can be chosen to make one error probability very small only at the expense of making the other large, at least for values θ near 1. Classically, a value α between 0 and 1 was often specified, reflecting the maximum error probability that the experimenter was willing to tolerate under that one of the two hypotheses that was singled out as representing (for example) a more traditional or conservative scientific theory. Subject to this restriction, one would select a procedure that would perform as well as possible, in some well-defined sense, under the other hypothesis.

In the present example of uniformly distributed X_i, let us consider nonsequential procedures t_n based on $X_1, X_2, \ldots, X_n$, and suppose $\theta = 1$ is the hypothesis under which the probability of error is to be $\leqslant\alpha$. If we again write $V_n = \max_{1\leqslant i\leqslant n} X_i$, it can be shown that, among all procedures t_n for which the probability of error is $\leqslant \alpha$ when $\theta = 1$, the procedure t_n^* defined by

$$t_n^*(X_1, \ldots, X_n) = \begin{cases} g_0 & \text{if } V_n \leqslant (1 - \alpha)^{1/n}, \\ g_1 & \text{if } V_n > (1 - \alpha)^{1/n}, \end{cases} \tag{1.4}$$

yields the maximum probability of a correct guess when $\theta > 1$. Since, for $0 < c < \theta$,

$$P_\theta\{V_n \leqslant c\} = P_\theta\{X_1 \leqslant c, X_2 \leqslant c, \ldots, X_n \leqslant c\} = (c/\theta)^n, \tag{1.5}$$

the probability of an incorrect guess under t_n^* is

$$P_\theta\{t_n^* \text{ is incorrect}\} = \begin{cases} \alpha & \text{if } \theta = 1, \\ (1 - \alpha)/\theta^n & \text{if } \theta > 1. \end{cases} \tag{1.6}$$

(In this elementary setting, other procedures than that of (1.4) also yield (1.6); the one given by (1.4) is often used because it also performs optimally for true values $\theta < 1$ if such values are possible, where g_0 is also viewed as correct when $\theta \leqslant 1$.)

If α is a small number, such as 10^{-3}, scientists would often publish the fact that $t_n^* = g_1$ as fairly conclusive evidence that the true θ is not 1; for the small probability ($\leqslant .001$) of having the X_i unluckily turn out to yield g_1 when in fact $\theta = 1$ leads us to turn to belief in the alternative that $\theta > 1$.

Suppose a dishonest scientist (perhaps a consultant to the manufacturer of a stultifying aerospray the public will buy if it can be shown $\theta > 1$) wants to be sure to publish the conclusion g_1 using t_n^*, with a small α such as .001, *even if* $\theta = 1$. He can achieve this by using a sequential procedure while pretending he is not doing so, as follows: Let $N\#$ be the first integer n for which $V_n > (1 - \alpha)^{1/n}$, if such an n exists; if no such n exists with positive probability for some θ value, we shall say $N\#$ is *not well defined* for that θ value. If $N\# = n$, the scientist pretends he had decided on the sample size n and procedure t_n^* from the outset, and according to (1.6) can thus publish the statement indicated at the end of the previous paragraph. But actually he is using a sequential procedure based on $N\#$, and as long as

$$P_\theta\{N\# \text{ is well defined}\} = 1 \qquad \text{for } \theta \geqslant 1 \tag{1.7}$$

the resulting procedure $t_N^*\#$ achieves

$$P_\theta\{t_N^*\# \text{ is incorrect}\} = \begin{cases} 1 & \text{if } \theta = 1, \\ 0 & \text{if } \theta > 1, \end{cases} \tag{1.8}$$

in contrast to the claim (1.6) he is making.

We shall verify (1.7) when $\theta = 1$, it being more trivial when $\theta > 1$. For $i \geqslant 1$, let A_i be the event that $N\# \leqslant 2^i$, and let A_i^c be its complement. Clearly A_{i+1}^c entails

$$\max_{2^i < j \leq 2^{i+1}} X_j \leqslant (1-\alpha)^{1/2^{i+1}},$$

and since this last event is independent of $A_1, \ldots, A_i$, we have

$$P_{\theta=1}\{A_{i+1}^c \mid A_1^c \cap A_2^c \cap \cdots \cap A_i^c\} \leqslant P_{\theta=1}\{\max\nolimits_{2^i < j \leq 2^{i+1}} X_j$$
$$\leqslant (1-\alpha)^{1/2^{i+1}}\} = (1-\alpha)^{1/2}. \tag{1.9}$$

Since the sets A_i^c are decreasing, we obtain

$$P_{\theta=1}\{A_k^c\} = P\left\{\bigcap_{i=1}^{k} A_i^c\right\}$$
$$= P\{A_1^c\} \prod_{i=1}^{k-1} P\left\{A_{i+1}^c \,\middle|\, \bigcap_{j=1}^{i} A_j^c\right\} \leqslant (1-\alpha)^{(k+1)/2}, \tag{1.10}$$

from which $P_{\theta=1}\{A_k\} \to 1$ as $k \to \infty$, so that (1.7) is satisfied.

2. RANDOM WALKS AND MARTINGALES

Throughout this section we shall restrict consideration to discrete random variables X_i, Y_i, Z_i, in order to dispense with the measure-theoretic details that are inessential to the concepts of interest here.

The analysis of sequential procedures for many models that are less simple than those of Section 1, can be made in terms of certain chance processes we now discuss. Let $Y_1, Y_2, \ldots,$ be i.i.d. r.v.'s. Define $S_0 = 0$ and $S_n = \sum_1^n Y_i$ for $n \geqslant 1$. The process $S_0, S_1, S_2, \ldots,$ is called a *random walk* (more precisely, a one-dimensional random walk with discrete time, and which is homogeneous in space and time: the law of the motion $S_{n+1} - S_n$, from "time" n to $n+1$, is independent of the position S_n and of the time n). The possible values of the sequence $\omega = (S_0, S_1, S_2, \ldots)$ can be thought of as points in the probability space of the random walk, and it is often convenient to join successive pairs (n, s_n), $n \geqslant 0$, to make a piecewise-linear graph that can be viewed as the "sample path" graph of the random function ω. In the next section we will consider such random walks with "absorbing sets" defined by two sequences $\{a_n, n > 0\}$ and $\{b_n, n > 0\}$ with $b_n \leqslant a_n$; the stopping variable N is defined as $N = \min\{n : S_n \leqslant b_n \text{ or } S_n \geqslant a_n\}$, and in physical applications one thinks of the "particle" whose path is given by ω as being "absorbed" by the set $[a_N, +\infty)$ or $(-\infty, b_N]$ into which it falls when the process is stopped (Fig. 1).

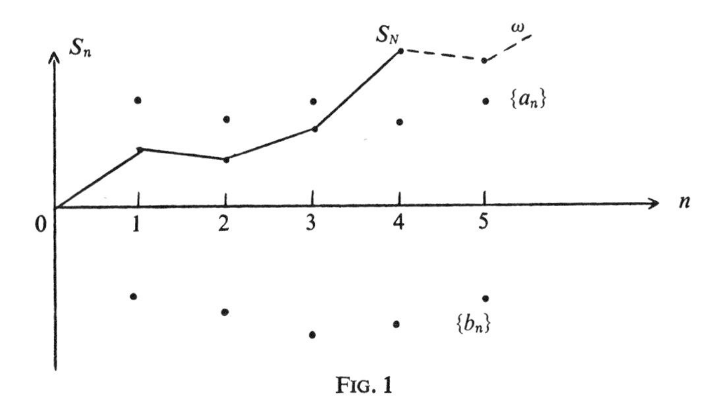

FIG. 1

If $EY_i = 0$, we see that, for any sequence of possible values $(0, s_1, \ldots, s_n)$ of $(S_0, S_1, \ldots, S_n)$, we have

$$E\{S_{n+1} \mid S_0 = 0, S_1 = s_1, \ldots, S_n = s_n\} = s_n$$

for the conditional expectation of the position S_{n+1} given the "past history." This means that the process $\{S_n\}$ is a special case of a 0-expectation *martingale*, a chance process $\{Z_n, n \geqslant 0\}$ with the properties $Z_0 = 0$, $E|Z_n| < \infty \forall n$, and

$$E\{Z_{n+1} \mid Z_0 = 0, Z_1 = z_1, \ldots, Z_n = z_n\} = z_n \tag{2.1}$$

for all n and any possible realization $(0, z_1, \ldots, z_n)$ of the past. This is more general than the 0-expectation random walk because the

$Z_{n+1} - Z_n$ do not have to be i.i.d. It is an immediate consequence of (2.1) that $EZ_n = 0 \forall n$.

Martingales are a mathematical model of "fair games of chance." If you play a game repeatedly and Z_n is your total "profit" (negative profit being loss) after n plays of the game, the game may be thought to be fair if, no matter what the history of your successes through n plays of the game, the conditional expectation of the amount $Z_{n+1} - Z_n$ you win on the $(n + 1)$st play is 0. That is precisely (2.1).

Everyone wants a strategy that will let him quit while ahead. A *stopping time* N based on $\{Z_i\}$ is a positive integer-valued random variable with the property that, for each positive integer n, the event $\{N = n\}$ is a set in the space of $Z_1, Z_2, \ldots, Z_n$. In the gambling context, this formalizes the notion that you can not in practice decide whether to stop gambling at time 10 on the basis of knowledge of whether or not the 11th play is to be favorable for you.

Can we obtain a favorable stopping strategy N, one for which $EZ_N > 0$, for any simple games? If N is bounded above, it is not hard to use (2.1) and to work backward from that upper bound (as $n + 1$), step by step, to show $EZ_N = 0$. On the other hand, in the simple game of repeatedly flipping a fair coin (with independent flips) while betting an adversary one unit on each flip that you will guess the outcome correctly, so that the amount you win on the nth flip (regardless of how you guess) is $Z_{n+1} - Z_n = Y_n$ where the Y_i are independent and $P\{Y_i = 1\} = P\{Y_i = -1\} = 1/2$, a well-known strategy is to "continue betting until you're 1 unit ahead, and then stop." In other words, $N = \min\{n : Z_n = 1\}$, in which case $Z_N = 1$ with probability one, and hence $EZ_N = 1$. The only feature of this development that is not obvious is that N is well defined with probability one, and this can be proved by an analogue of the proof of (1.7): defining $n_1 = 2$ and $n_{i+1} - n_i = n_i^2$, it can be shown, in analogy with (1.9), that $P\{Z_{n_{i+1}} \geqslant 1 \mid Z_{n_i} = z\} > \epsilon$ for all $i > 1$ and all $z \geqslant -n_i$ (the minimum possible value), for some $\epsilon > 0$.

However, a "practical" difficulty of this strategy is that it requires infinite capital to use it, since Z_n can be arbitrarily large and negative for some n prior to N. A theoretical difficulty of the strategy is that, although $P\{N < \infty\} = 1$, it can be shown that the probability law of N has such large "tails" that $EN = \sum_{n=1}^{\infty} nP\{N = n\} = +\infty$. Thus, although the gambler with infinite capital would eventually win with this strategy, the expected length of time it would take him to do it is infinite.

This last difficulty can be eliminated by using a different betting strategy, if your opponent will permit it: instead of betting one unit on each flip of the coin, you double your bet until you win. This means $Z_{n+1} - Z_n = \pm 2^n$, each possibility with probability 1/2, for $n \geqslant 0$, and again that $N = \min\{n : Z_n = 1\}$ and hence $Z_N = 1$ with probability 1. In this case $N = n$ if you lose the first $n - 1$ matches and win the nth, an event of probability 2^{-n}. This decreases rapidly, and $EN = \sum_{n=1}^{\infty} n2^{-n} = 2$ (see (1.2)), so the shortcoming of the previous strategy has been overcome. Alas, it has been replaced by another: the *expected amount you bet* on the final play is $E|Z_N - Z_{N-1}| = E2^{N-1} = \sum_{n=1}^{\infty} 2^{n-1}2^{-n} = +\infty$. Thus, the "practical" difficulty of the previous paragraph is even more evident here.

The peculiarities of these two strategies typify what one must resort to in order to obtain a positive expected gain from imposing a stopping time on a fair game. Doob's *optional stopping theorem* for martingales makes precise the impossibility of designing a well-behaved winning strategy, by asserting that, if the laws of N and $Z_{n+1} - Z_n$ are suitably regular, $EZ_N = 0$. The result has an especially simple statement in the case of the random walk $Z_n = S_n$:

$$EY_i = 0 \quad \text{and} \quad EN < \infty \quad \text{imply} \quad ES_N = 0. \tag{2.2}$$

This last is a version of what is known as *Wald's equation*, and it can be proved by using the validity of (2.2) for the previously mentioned case of a bounded stopping variable $N_m = \min(N, m)$ and letting $m \to \infty$. We shall use (2.2) in the next section.

3. WALD'S SEQUENTIAL PROBABILITY RATIO TEST

Suppose $X_1, X_2, \ldots,$ are i.i.d. r.v.'s, it being known that the common probability function of the X_i is one of two different specified laws f_0 and f_1, where $f_j(x) = P_j\{X_i = x\}$; it is unknown whether f_0 or f_1 is the true law, and the problem, in the language of Section 1, is to decide between these two hypotheses. Again we illustrate the ideas in the discrete case, and for convenience we assume f_0 and f_1 positive on the same domain.

We define the r.v. $\lambda_n = \prod_{i=1}^n [f_1(X_i)/f_0(X_i)]$, the "probability ratio" of the observations under the two possible laws. Intuitively, if $X_1, \ldots, X_n$ turn out such as to make λ_n large, one might feel secure in making the guess g_1 that f_1 is the true law, and if λ_n is small one would vote for f_0. This notion is made precise in the Neyman-Pearson Lemma, according to which the nonsequential procedure

$$t_n^*(X_1, \ldots, X_n) = \begin{cases} g_0 & \text{if } \lambda_n < c, \\ g_1 & \text{if } \lambda_n \geqslant c, \end{cases} \tag{3.1}$$

where c is a nonnegative constant, has the property that, if t_n is any other procedure, and

$$P_0\{t_n \neq g_0\} \leqslant P_0\{t_n^* \neq g_0\}, \tag{3.2}$$

then

$$P_1\{t_n \neq g_1\} \geqslant P_1\{t_n^* \neq g_1\}. \tag{3.3}$$

Thus, as indicated in connection with (1.4) (which conforms with (3.1) in an extreme example), no procedure can simultaneously minimize both probabilities of error $P_i\{t_n \neq g_i\}$, but for each value of c the corresponding t_n^* has the property that no t_n can be better than it in terms of both of these error probabilities. Alternatively, in terms of the language used in Section 1, if c has been chosen in such a way as to make $P_0\{t_n^* \neq g_0\}$ equal a specified value α, then among all t_n with $P_0\{t_n \neq g_0\} \leqslant \alpha$, the procedure t_n^* minimizes $P_1\{t_n \neq g_1\}$. The Neyman-Pearson Lemma is not difficult to prove, but we shall not take the space to prove it here.

Intuitively, large or small values of λ_n seem more conclusive than values near 1. This suggests the possible wisdom of using a sequential procedure that continues to observe the X_i until the first time $N = n$ that λ_n is "suitably large or small." This last means choosing two sequences A_n and B_n with $0 < B_n < A_n < \infty$ and stopping at $N = \min\{n : \lambda_n \leqslant B_n \text{ or } \lambda_n \geqslant A_n\}$. The simplest such sequence is obtained by taking A_n and B_n to be constants A and B, independent of n, and which satisfy $0 < B < 1 < A < \infty$. This is the form of Wald's *sequential probability ratio test:*

$$\bar{N} = \min\{n : \lambda_n \leqslant B \text{ or } \lambda_n \geqslant A\},$$
$$\bar{t}_{\bar{N}} = \begin{cases} g_0 & \text{if } \lambda_{\bar{N}} \leqslant B, \\ g_1 & \text{if } \lambda_{\bar{N}} \geqslant A. \end{cases} \tag{3.4}$$

This procedure takes advantage of "lucky observation sequences" as described in Section 1, to stop at an early time n if λ_n is far from 1 then.

The simple form of this test enabled Wald to compute good approximations to the performance characteristics $P_i\{\bar{t}_{\bar{N}} \neq g_i\}$ and $E_i\bar{N}$ discussed in the introduction, and also suggested an optimum property of this test among all sequential (and nonsequential) tests. We shall discuss these in this section and will also give a simple numerical example. First we note that, if we write $Y_i = \log[f_1(X_i)/f_0(X_i)]$, the Y_i are i.i.d. r.v.'s under either f_0 or f_1 (as the law of X_i), and in the notation of the random walk of Section 2 we can rewrite (3.4) as

$$\bar{N} = \min\{n : S_n \leqslant \log B \text{ or } S_n \geqslant \log A\}$$
$$\bar{t}_{\bar{N}} = \begin{cases} g_0 & \text{if } S_{\bar{N}} \leqslant \log B, \\ g_1 & \text{if } S_{\bar{N}} \geqslant \log A. \end{cases} \tag{3.5}$$

We can thus think of Wald's test in terms of a random walk with one of two laws and with the absorbing sets of Figure 1 defined by $a_n = \log A$ and $b_n = \log B$. That, in accordance with our intuition, the random walk tends to drift toward the upper barrier $\log A$ if f_1 is true and toward the lower barrier $\log B$ if f_0 is true, is indicated by the fact that

$$E_0 Y_1 < 0 < E_1 Y_1, \tag{3.6}$$

which is a consequence of Jensen's inequality: since $\log y$ is concave, the average of its value at several points is less than its value at the average point. Hence,

$$\begin{aligned} E_0 Y_1 &= E_0 \log[f_1(X_1)/f_0(X_1)] < \log E_0[f_1(X_1)/f_0(X_1)] \\ &= \log \sum_x [f_1(x)/f_0(x)]f_0(x) = \log 1 = 0, \end{aligned} \tag{3.7}$$

the strictness of the inequality following from the fact that there is more than one possible value of Y_1 since the f_i are different. The other half of (3.6) is proved similarly.

For any well-defined N and associated guessing procedure t_N, we hereafter abbreviate $P_i\{t_N \neq g_i\}$ by $\pi_i(t_N)$.

It is not hard to verify that $\bar{N}$ is well defined and that $E_i\bar{N} < \infty$. Simple inequalities on the probabilities of error for $\bar{t}_{\bar{N}}$ are

$$\pi_0(\bar{t}_{\bar{N}})/[1 - \pi_1(\bar{t}_{\bar{N}})] \leqslant A^{-1}; \qquad \pi_1(\bar{t}_{\bar{N}})/[1 - \pi_0(\bar{t}_{\bar{N}})] \leqslant B. \quad (3.8)$$

The first of these, for example, is proved by noting that, if $\bar{Q}_n = \{(x_1, \ldots, x_n) : \bar{N} = n, \bar{t}_{\bar{N}} = g_1\}$, we have $\lambda_n \geqslant A$ on $\bar{Q}_n$, and thus

$$\begin{aligned} \pi_0(\bar{t}_{\bar{N}}) &= \sum_{n=1}^{\infty} P_0\{\bar{Q}_n\} = \sum_{n=1}^{\infty} \sum_{\bar{Q}_n} \prod_{i=1}^{n} f_0(x_i) \\ &\leqslant \sum_{n=1}^{\infty} \sum_{\bar{Q}_n} A^{-1} \prod_{i=1}^{n} f_1(x_i) = A^{-1} \sum_{n=1}^{\infty} P_1\{\bar{Q}_n\} \\ &= A^{-1}[1 - \pi_1(\bar{t}_{\bar{N}})]. \end{aligned} \quad (3.9)$$

The inequality in (3.9) arises because $S_{\bar{N}}$ might be $> \log A$ rather than exactly $\log A$, on $\bar{Q}_n$. If $S_{\bar{N}}$ can only take on the values $\log A$ and $\log B$, the inequalities (3.8) become equalities, and we obtain $\pi_0(\bar{t}_N) = (1 - B)/(A - B)$ and $\pi_1(\bar{t}_N) = B(A - 1)/(A - B)$. In practice, even when $S_{\bar{N}}$ is not of this special character, these formulas usually give good approximations.

We now turn to a computation of a lower bound on E_iN for any well-defined stopping time N for which $E_iN < \infty$ and $\pi_i(t_N) < 1$ for both i. First, writing $Q_n = \{(x_1, \ldots, x_n) : N = n, t_N = g_1\}$, we compute in a manner similar to (3.9),

$$\begin{aligned} E_1\{1/\lambda_N \mid t_N = g_1\} &= \sum_n \sum_{Q_n} \left[\prod_1^n (f_0(x_i)/f_1(x_i))\right] \prod_1^n f_1(x_i)/P_1\{t_N = g_1\} \\ &= \pi_0(t_N)/[1 - \pi_1(t_N)]. \end{aligned} \quad (3.10)$$

Next, using Jensen's inequality as in (3.7), with (3.10), we obtain

$$\begin{aligned} E_1\{S_N \mid t_N = g_1\} &= -E_1\{\log(1/\lambda_N) \mid t_N = g_1\} \geqslant -\log E_1\{1/\lambda_N \mid t_N = g_1\} \\ &= \log\{[1 - \pi_1(t_N)]/\pi_0(t_N)\}. \end{aligned} \quad (3.11)$$

A similar computation yields

$$E_1\{S_N \mid t_N = g_0\} \geqslant \log\{\pi_1(t_N)/[1 - \pi_0(t_N)]\}.$$

Together these give

$$\begin{aligned} E_1 S_N &= \sum_{i=0}^{1} E_1\{S_N \mid t_N = g_i\} P_1\{t_N = g_i\} \\ &\geqslant \pi_1(t_N) \log\{\pi_1(t_N)/[1 - \pi_0(t_N)]\} \\ &\quad + [1 - \pi_1(t_N)] \log\{[1 - \pi_1(t_N)]/\pi_0(t_N)\} \\ &= q(\pi_0(t_N), \pi_1(t_N)) \text{ (say).} \end{aligned} \tag{3.12}$$

Defining Z_n as the sum $\sum_1^n (Y_i - E_1 Y_i)$, still with

$$Y_i = \log[f_1(X_i)/f_0(X_i)],$$

we may apply Wald's equation (2.2) to obtain $E_1 S_N = (E_1 Y_1) E_1 N$ and hence, from (3.12) and (3.6),

$$E_1 N \geqslant (E_1 Y_1)^{-1} q(\pi_0(t_N), \pi_1(t_N)). \tag{3.13}$$

Similarly, one can prove that

$$E_0 N \geqslant (-E_0 Y_1)^{-1} q(\pi_1(t_N), \pi_0(t_N)). \tag{3.14}$$

We note that the inequality sign in (3.12), and thus in (3.13) or (3.14), becomes equality if S_N can take on only one value when $t_N = g_1$ and only one when $t_N = g_0$. For Wald's $\bar{N}$, this means that if $S_{\bar{N}}$ can only take on the values $\log B$ and $\log A$, the right sides of (3.13) and (3.14) give exact expressions for $E_i\bar{N}$. As in the case of (3.8), even when $S_{\bar{N}}$ is not of this special character, the expressions of (3.13) and (3.14) often give excellent approximations for Wald's stopping rule, in practice.

Wald conjectured that the simple sequential rule (3.4) possesses a very strong optimum character, compared with all others. Suppose we specify positive values α_0 and α_1, α_j being the maximum probability of an incorrect guess that we are willing to tolerate when f_j is the true law of the X_i. Subject to the two restrictions $\pi_i(t_N) \leqslant \alpha_i$ $(i = 0, 1)$, we might seek the procedure (N, t_N) that minimizes $E_0 N$; or we might try to find the procedure that minimizes $E_1 N$. In mathematics one may expect two different minimization problems to have different solutions; the problems are concerned with achieving quick stopping time on the average under two quite different drifts, as indicated in (3.6). However, remarkably, if there is a Wald procedure (3.4) for which $\pi_i(\bar{t}_{\bar{N}}) = \alpha_i$ for $i = 0, 1$ (which there always is if $\alpha_0 + \alpha_1 < 1$ and Y_1 has a continuous law under both f_i), then that procedure is simultaneously the solution to both problems! This is stated as:

Optimum character of Wald's test. *If* $(\bar{N}, \bar{t}_{\bar{N}})$ *is a procedure of the form* (3.4) *and* (N, t_N) *is any other procedure for which*

$$\pi_i(t_N) \leqslant \pi_i(\bar{t}_{\bar{N}}), \qquad i = 0, 1, \tag{3.15}$$

then

$$E_i N \geqslant E_i \bar{N}, \qquad i = 0, 1. \tag{3.16}$$

Wald proved this in his book assuming $E_i N < \infty$ and that $S_{\bar{N}}$ can only equal $\log A$ or $\log B$. This last assumption is very restrictive, and the optimum character in the general setting without this assumption was first proved by Wald and Wolfowitz (1948). The proof assuming that $S_{\bar{N}}$ can only equal $\log A$ or $\log B$ and that the $E_i N$ are finite is quite easy: Differentiation shows that $q(\pi_0, \pi_1)$ is decreasing in its arguments provided $\pi_0 + \pi_1 < 1$, a condition always satisfied for a Wald procedure. Hence, from (3.15), the expressions on the right sides of (3.13) and (3.14) are always at least equal to the corresponding expressions with the $\pi_i(\bar{t}_{\bar{N}})$ in place of the $\pi_i(t_N)$. On the other hand, our earlier remarks indicated that these last expressions are equal to the $E_i\bar{N}$ for $S_{\bar{N}}$ of the assumed special character. While this proof does not carry over to the general case, it makes the conclusion there plausible.

To illustrate the savings in $E_i N$ that are possible from using a Wald sequential test rather than a nonsequential one, we consider the problem of deciding whether a coin has probability .4 or .6 of coming up "heads" on a single toss, assuming those are known to be the only possible values of that probability. If $X_i = 1$ or -1 corresponding to a head or tail, this means that

$$f_0(-1) = .6 = 1 - f_0(1), \qquad f_1(-1) = .4 = 1 - f_1(1), \tag{3.17}$$

with the $f_i(x) = 0$ otherwise. Consequently, Y_i can only take on the values $\pm\log 1.5$, with

$$P_1\{Y_i = \log 1.5\} = .6, \qquad P_0\{Y_i = \log 1.5\} = .4. \qquad (3.18)$$

If, for example, we specify $\alpha_0 = \alpha_1 = 1/25$ as the maximum probability of error we will tolerate under either f_0 or f_1, binomial probability function tables show that 75 observations are needed for a nonsequential test to achieve this, the optimum (Neyman-Pearson) procedure being $t_{75} = g_1 \Leftrightarrow S_{75} > 0$ (which yields $\pi_i(t_{25}) = .0396$). For a sequential Wald test, we take $\log A = -\log B$ to be of the form $m \log 1.5$ where m is an integer. Then $S_{\bar{N}}$ can only equal $\log A$ or $\log B$, and the discussion below (3.9) shows that, in order to achieve $\pi_0(\bar{t}_{\bar{N}}) = \pi_1(\bar{t}_{\bar{N}}) \leqslant 1/25$, we need only solve $1/25 \geqslant (1 - B)/(A - B) = 1/(A + 1)$, for which we only need $m = 8$, the smallest integer $\geqslant \log 24/\log 1.5$ (which yields $\pi_i(\bar{t}_{\bar{N}}) = .0376$). The expressions (3.13) and (3.14) are equalities for the Wald test, and (since (3.18) yields $E_1 Y_1 = .2 \log 1.5$) they yield $E_i\bar{N} = 37.0$. Thus, the expected number of observations required by the Wald test is slightly *less than half* that required by the best nonsequential test. This is typical of what occurs in more complex settings.

4. SEQUENTIAL PROCEDURES FOR OTHER PROBLEMS

The striking success of sequential analysis in Section 3 was presented there for the simple hypothesis testing problem in which there was only one possible probability law of the X_i under each hypothesis. In this section we shall mention briefly a few of the many more complicated statistical settings which have been handled with sequential procedures. For simplicity, for the most part we consider once more flips of a coin, but no longer with the probability of a head being restricted to two possible values as in the example of Section 3. Thus, we consider i.i.d. "Bernoulli r.v.'s" X_i with

$$P_p\{X_i = 1\} = 1 - P_p\{X_i = -1\} = p \qquad (4.1)$$

for $0 < p < 1$.

(A) First suppose, in extension of the hypothesis testing problem of Section 3, that we want to guess whether the true but unknown value of p that governs this model satisfies $0 < p < 1/2$ or $1/2 < p < 1$—do we guess that the coin is biased in favor of tails (g_0) or in favor of heads (g_1)? A moment's reflection shows that, since the law of X_i is almost the same when $p = 1/2 - \epsilon$ as when $p = 1/2 + \epsilon$ when ϵ is small, no nonsequential procedure can achieve

$$\sup_{p \neq 1/2} P_p\{t_N \text{ is incorrect}\} \leqslant \alpha_0, \tag{4.2}$$

no matter how large the fixed sample size $N = n$ is, if $\alpha_0 < 1/2$. Similarly, the requirement (4.2) cannot be achieved by the sequential procedure that stops the first time $\sum_1^n X_i$ is $\leqslant C$ or $\geqslant D$ (this being the form that (3.5) takes for testing whether $p = 1/2 - \epsilon_0$ or $p = 1/2 + \epsilon_0$, for a *fixed* ϵ_0). The simplest type of procedure that will achieve (4.2) is based on a stopping rule $N' = \min\{n : |\sum_1^n X_i| \geqslant a_n\}$, where $\{a_n\}$ is a sequence of positive constants for which

$$\sup_{p<1/2} P_p\left\{\sup_{n>0}\left(\sum_1^n X_i - a_n\right) \geqslant 0\right\} \leqslant \alpha_0 \tag{4.3}$$

and

$$P_p\left\{\inf_{n>0}\left(\sum_1^n X_i + a_n\right) < 0\right\} = 1, \qquad 0 < p < 1/2. \tag{4.4}$$

If we use such an N' with $t_{N'} = g_0$ or g_1 according to whether $\sum_1^{N'} X_i < 0$ or > 0, equation (4.4) insures that N' is well defined, since, for $p < 1/2$, the random walk $\sum_1^n X_i$ eventually hits or crosses the lower barrier $\{-a_n, n > 0\}$, if it does not cross the upper barrier first. On the other hand, (4.3) insures that, when $p < 1/2$, the probability that the random walk *ever* hits the upper barrier $\{a_n, n > 0\}$, which is greater than the probability that it results in the incorrect guess g_1 due to hitting the upper barrier before the lower one, is $\leqslant \alpha_0$. The case $p > 1/2$ is covered similarly.

The condition (4.4) follows if the a_n satisfy

$$\lim_{n\to\infty} a_n/n = 0, \tag{4.5}$$

since the well-known law of large numbers asserts for these X_i with $E_p X_i = 2p - 1$, that

$$\lim_{n\to\infty} P_p\left\{\left|n^{-1}\sum_1^n X_i - (2p-1)\right| < \epsilon\right\} = 1 \qquad (4.6)$$

for every $\epsilon > 0$; choosing ϵ to be $< 1 - 2p$, we obtain (4.4). In order to study (4.3), we first note the intuitively plausible statement that the probability of (4.3) is nondecreasing in p. This statement can be proved by constructing r.v.'s X_i' and X_i'' as functions of the same r.v. H_i uniformly distributed on (0, 1), with $X_i' = 1$ or $= -1$ depending on whether $H_i \leqslant p'$ or $> p'$, and with $X_i'' = 1$ or -1 depending on whether $H_i \leqslant p''$ or $> p''$; then X_i' and X_i'' have the law (4.1) with $p = p'$ and p'', and if $p' < p''$ we have $X_i'' \geqslant X_i'$ and thus $\sum_1^n X_i'' \geqslant \sum_1^n X_i'$, from which the monotonicity of (4.3) follows. As a consequence of this monotonicity, (4.3) follows from

$$P_{1/2}\left\{\sup_{n>0}\left(\sum_1^n X_i - a_n\right) \geqslant 0\right\} \leqslant \alpha_0. \qquad (4.7)$$

(We use the law (4.1) with $p = 1/2$ in (4.7), even though it is not considered to be a possible law for the actual coin at hand.) There are many possible sequences $\{a_n\}$ that satisfy (4.5) and (4.7). From our earlier discussion we know that a_n cannot be chosen constant, and we seek unbounded a_n. Elementary probability inequalities on the tails of the binomial probability law (e.g., in Feller (1950)) show that $a_n = Cn^\beta$, with $1/2 < \beta < 1$ and C sufficiently large, suffices; this is reflected in the fact that Cn^β is $Cn^{\beta - 1/2}$ "standard deviation units" of the law of $\sum_1^n X_i$. It is thus not hard to show that

$$\sum_{n=1}^{\infty} P_{1/2}\left\{\sum_1^n X_i \geqslant a_n - 1\right\} \leqslant \alpha_0$$

for C sufficiently large, and this sum dominates the probability of (4.7). In order that $E_p N'$ not be too large when p is near 1/2 (the most difficult values of p to judge correctly), we want the a_n to increase as slowly as possible with n subject to (4.7), and this suggests letting β be near 1/2. In fact, it can be shown that $Cn^{1/2} \log n$ also works for appropriate C. A more subtle argument is involved in showing that, for each $\delta > 0$, the sequence

$$C + [(2 + \delta)n \log\log(n + 3)]^{1/2}$$

works for sufficiently large C. In a sense this is the smallest type of regular sequence that works, in that for $\delta < 0$ it can be shown that the probability of (4.7) is 1, and that the left side of (4.2) equals 1/2 for that sequence. The properties of these borderline sequences are the subject of the celebrated "Law of the Iterated Logarithm" of probability theory. (See Feller (1950).)

The property (4.7) implies that, when $p = 1/2$, the r.v. N' is not well defined, since with probability at least $1 - 2\alpha_0$ the rule never tells us to stop! This is not merely a property of rules constructed through the use of (4.3)–(4.4), and it can be shown that every stopping rule that satisfies (4.2) has this property. We excluded $p = 1/2$ from our set of possible values governing the law of the X_i, but in our next example we use this property of the value $p = 1/2$ (and a corresponding one for $p > 1/2$).

(B) Our second problem is to find a procedure that still achieves (4.2) for $p \geqslant 1/2$, but that demands *no error* if $p < 1/2$; that is, we require

$$P_p\{t_N \text{ incorrect}\}\begin{cases} \leqslant\alpha_0 & \text{if } p \geqslant 1/2, \\ =0 & \text{if } p < 1/2. \end{cases} \tag{4.8a}$$

The idea of considering such a requirement is due to Robbins. It is impossible to achieve (4.8a) if N is well defined for all p, and the essence of the procedure is that it makes use of the fact that $P_p\{\text{the procedure never stops}\} \geqslant 1 - \alpha_0$ for $p \geqslant 1/2$ to assert the first half of (4.8a). The second half of (4.8a) can be written in the stronger form $P_p\{N \text{ well defined and } t_N \text{ correct}\} = 1$ for $p < 1/2$. A procedure that achieves these results is obtained by using only the lower boundary $\{-a_n\}$ of the previous problem: now $N = \min\{n : \sum_1^n X_i \leqslant -a_n\}$, and otherwise N is undefined. We guess g_0 if we stop. Then (4.4) yields the second half of (4.8a), and the first half follows from the analogue for $p > 1/2$ of (4.3) and (4.7):

$$\sup_{p\geq 1/2} P_p\left\{\inf_{n>0}\left(\sum_1^n X_i + a_n\right) \leqslant 0\right\} = P_{1/2}\left\{\inf_{n>0}\left(\sum_1^n X_i + a_n\right) \leqslant 0\right\} \leqslant \alpha_0. \tag{4.8b}$$

One may imagine applications of such a test: a physical process of some kind is "in control" if $p \geqslant 1/2$, and out of control (hence, requiring action) if $p < 1/2$. In the latter case, we want definitely to stop and "guess g_0." In the former case, we keep observing except for the probability α_0 that we incorrectly stop and go to the trouble of attempting to fix the process when it is actually under control. Obviously, the sense in using such a procedure will depend on the cost of taking many observations, perhaps of never stopping, compared with the possible cost of not detecting that the process is out of control when $p < 1/2$.

In certain extreme (non-Bernoulli) models for the law of the X_i's (replacing (4.1)) it is possible to find a sequential test that achieves $P_\theta\{N$ well defined and t_N correct$\} = 0$ for all true values θ of the parameter value, even though the two sets of values $\{\theta : g_i$ is a correct guess$\}$ represent collections of laws with a common limit point (as with $p = 1/2$ in the example just above). For example, if X_i has uniform density from $\theta - 1/2$ to $\theta + 1/2$, for deciding whether $\theta < 0$ or $\theta > 0$ we need only stop at $N = \min\{n : |X_n| > 1/2\}$ and make the obvious guess.

(C) For a final problem requiring the use of sequential analysis, we consider estimation of the parameter p of the law (4.1) of the X_i where $0 < p < 1$ and p is otherwise unknown. If, for some specified value c with $0 < c < 1$, we regard a guess (estimate) of the true value of p as correct if it differs by at most c units from p, and want an estimator t_n that is incorrect with probability at most α_0 (specified), there is no difficulty in finding an n for which the nonsequential procedure based on n observations, and which estimates p by

$$\begin{aligned} t_n(X_1, \ldots, X_n) &= \Big(n^{-1}\sum_1^n X_i + 1\Big)/2 \\ &= n^{-1}\,(\text{number of heads in } n \text{ tosses}), \end{aligned} \tag{4.9}$$

achieves this goal. However, suppose instead of wanting the *absolute error* $|t_n - p| \leqslant c$ as above that we are interested in making the *relative error* small, so that we want, for specified positive $c' < 1$ and α_0' (with $0 < \alpha_0' < 1$),

$$\sup_{0<p<1} P_p\{|t_n(X_1, \ldots, X_n) - p|/p \geqslant c'\} \leqslant \alpha_0'. \tag{4.10}$$

We now show that no fixed sample size n suffices. For, whatever value n is chosen, it is clear from (4.1) that

$$\lim_{p \downarrow 0} P_p\{X_1 = X_2 = \cdots = X_n = -1\} = 1.$$

Hence, whatever value $t_n(-1, -1, \ldots, -1) = h$ (say) may be, the probability approaches 1 as $p \downarrow 0$, that $|t_n - p|/p = |h/p - 1|$, which is either unbounded (if $h \neq 0$) or 1 (if $h = 0$ is permitted as a guess, even though it is not a possible value of p). In either case, (4.10) is violated.

The difficulty here occurs for p near 0. If there were a known value $r > 0$ and we *knew* $r \leqslant p < 1$, it would be easy to solve the problem with a nonsequential procedure. For example, suppose with $c = c'r$ that $\nu(r)$ is an integer such that taking $n = \nu(r)$ observations solves the original "absolute error" problem of the previous paragraph. Then, for $p \geqslant r$, we have

$$\alpha_0 \geqslant P_p\{|t_{\nu(r)} - p| \geqslant c'r\} \geqslant P_p\{|t_{\nu(r)} - p|/p \geqslant c'\}, \tag{4.11}$$

in accordance with (4.10) for $\alpha_0 = \alpha_0'$. The trouble, of course, is that we do not know such an r. Suppose we could find a stopping time N_1 (well defined for $0 < p < 1$) and a positive r.v. $R_{N_1}(X_1, \ldots, X_{N_1})$, which is "usually right" as a guess of such an r, in the sense that

$$P_p\{R_{N_1} > p\} \leqslant \alpha_0'/2 \qquad \text{for } 0 < p < 1. \tag{4.12}$$

We could then proceed as follows: (1) Observe $X_1, X_2, \ldots, X_{N_1}$, and compute R_{N_1}; (2) take $\nu^*(R_{N_1})$ *additional* observations, where $\nu^*(r)$ is the function ν that yields (4.11) for $p \geqslant r$ when $\alpha_0 = \alpha_0'/2$; (3) Use $t_{\nu^*(R_{N_1})}(X_{N_1+1}, \ldots, X_{N_1+\nu^*(R_{N_1})}) = t^*$ (say) to estimate p as in (4.9), but using only the additional observations of (2). The total number of observations is $N_1 + \nu^*(R_{N_1})$, the final stopping time. We then have, writing G_p for the d.f. of R_{N_1} when p is true,

$$\begin{aligned} P_p\{|t^* - p|/p \geqslant c'\} &\leqslant P_p\{R_{N_1} > p\} \\ &\quad + \int_0^p P_p\{|t^* - p|/p \geqslant c' | R_{N_1} = r \leqslant p\} dG_p(r) \\ &\leqslant \alpha_0'/2 + \alpha_0'/2 = \alpha_0', \end{aligned} \tag{4.13}$$

thus achieving (4.10).

It remains to define an N_1 and R_{N_1} that yield (4.12). Intuitively, the smaller p, the longer it will tend to take before any X_i is 1, so we might try to use $N_1 = \min\{n : X_n = 1\}$ as a stopping time, and also make R_{N_1} small when N_1 is large. For this N_1, we have $P_p\{N_1 > m\} = P_p\{X_1 = X_2 = \cdots = X_m = -1\} = (1 - p)^m$ when m is a nonnegative integer (from which it follows that N_1 is well defined for $0 < p < 1$), and consequently

$$P_p\{N_1 > u\} \geqslant (1 - p)^u \tag{4.14}$$

for all nonnegative u. Define

$$R_{N_1} = 1 - (1 - \alpha_0'/2)^{1/N_1}. \tag{4.15}$$

Then, for $0 < p < 1$,

$$\begin{aligned} P_p\{R_{N_1} \leqslant p\} &= P_p\{1 - (1 - \alpha_0'/2)^{1/N_1} \leqslant p\} \\ &= P_p\left\{N_1 \leqslant \frac{\log(1 - \alpha_0'/2)}{\log(1 - p)}\right\} \\ &= 1 - P_p\left\{N_1 > \frac{\log(1 - \alpha_0'/2)}{\log(1 - p)}\right\} \\ &\leqslant 1 - (1 - p)^{\log(1 - \alpha_0'/2)/\log(1 - p)} = \alpha_0'/2, \end{aligned} \tag{4.16}$$

where the inequality follows from (4.14). Thus, this N_1 and the R_{N_1} defined by (4.15) achieve (4.12).

The method used to solve this problem has many other applications. It amounts to first finding a (random) domain which is very likely to contain the true parameter value, and which is such that, *assuming* the true parameter value lies in that domain, we know a fixed sample size procedure (with sample size depending on the domain) that solves the problem. The idea of using such a method is originally due to Charles Stein.

There are many other statistical problems than those illustrated in this paper, in which only sequential procedures can yield solutions, or in which some sequential procedure is superior to any fixed sample size procedure. This is not always the case. For example, in the problem of estimating the unknown mean θ of i.i.d. normal r.v.'s X_i with known variance, it can be shown that, among all procedures with $\sup_{-\infty<\theta<\infty} E_\theta N \leq m_0$ (= specified integer), the maximum risk $\sup_{-\infty<\theta<\infty} P_\theta\{|t_N - \theta| \geqslant c\}$ is minimized by the fixed sample size procedure $N = m_0$, $t_{m_0} = m_0^{-1} \sum_1^{m_0} X_i$. (It can be made precise that there are no "lucky" observation sequences here since this t_n contains all the information about θ based on $X_1, \ldots, X_n$, and no value of t_n is more informative than another.) But such settings are really the rare ones, and sequential analysis often provides a valuable practical device as well as an interesting implementation of various probabilistic ideas.

SELECTED REFERENCES

W. Feller, *An Introduction to Probability Theory and Its Applications*, Wiley, New York, 1950 and thereafter.

Z. Govindarajulu, *Sequential Statistical Procedures*, Academic Press, New York, 1975.

J. Kiefer, "Invariance, sequential estimation, and continuous time processes," *Ann. Math. Statist.*, **28** (1957), 573–601.

H. Robbins, "Some aspects of the sequential design of experiments," *BAMS*, **58** (1952), 527–535.

———, "Statistical methods related to the law of the iterated logarithm," *Ann. Math. Statist.*, **41** (1970), 1397–1409.

C. Stein, "A two-sample test for a linear hypothesis whose power is independent of the variance," *Ann. Math. Statist.*, **16** (1945), 243–258.

A. Wald, *Sequential Analysis*, Wiley, New York, 1947.

A. Wald and J. Wolfowitz, "Optimum character of the sequential probability ratio test," *Ann. Math. Statist.*, **19** (1948), 326–339.

J. Wolfowitz, "Minimax estimates of the mean of a normal distribution with known variance," *Ann. Math. Statist.*, **21** (1950), 218–230.

DEPENDENCE AND ASYMPTOTIC INDEPENDENCE FOR RANDOM PROCESSES

*Murray Rosenblatt**

1. INTRODUCTION

Most of the classical results in probability theory are concerned with statistically independent events and observations. However, even in the past century it was clear that there were many situations in which some aspect of statistical dependence arose naturally and had to be dealt with. In a sense, the different notions of ergodicity and mixing that we shall deal with can be thought of as giving us a measure of deviation from independence. There is especial interest in finding conditions under which classical results can be extended so as to cover interesting types of dependence. First of all some basic concepts will be discussed. Gaussian and Markov processes are then

* Research noted supported in part by the Office of Naval Research.

introduced. The definitions of ergodicity and mixing together with strengthened versions like uniform ergodicity and strong mixing are presented. What these concepts mean for Gaussian and Markov processes is considered at some length. There is a discussion of the central limit theorem and some related results.

2. BASIC CONCEPTS

Consider a doubly infinite vector of real numbers $\omega = (x_n, -\infty < n < \infty)$ with nth coordinate $\{\omega\}_n = x_n = x_n(\omega)$. Think of this as representing a sequence of observations on some aspect of a system from the infinite past to the infinite future. Suppose we consider the space of all such vectors and special sets of vectors, for example,

$$\begin{aligned}\{\omega : x_{n_1}(\omega) \leqslant y_1, x_{n_2}(\omega) \leqslant y_2, \ldots, x_{n_k}(\omega) \leqslant y_k\}\\ = B(n_1, y_1; \ldots; n_k, y_k),\end{aligned}$$

the set of vectors with n_1st coordinate less than or equal to $y_1, \ldots,$ and n_kth coordinate less than or equal to y_k. Here $n_1 < n_2 < \cdots < n_k$ and the y's are real numbers. The coordinate functions or *random variables* $\{x_n(\omega), -\infty < n < \infty\}$ are statistically *independent* if the probability (or intuitively the relative likelihood) of the set or "event" $B(n_1, y_1; \ldots; n_k, y_k)$

$$P(B(n_1, y_1; \ldots; n_k, y_k)) = F(y_1)F(y_2) \cdots F(y_k) \tag{2.1}$$

takes the product form given in (2.1) with

$$F(y) = P(B(n, y)),$$

the one-dimensional *distribution function* of the random variables $\{x_n(\omega)\}$, that is, the joint probability distribution is the product of the marginal or one-dimensional distributions. The relation (2.1) is assumed to hold for all integers $k \geqslant 1$ and all real y's. For simplicity it has been assumed that the $x_j(\omega)$'s all have the same one-dimensional distribution. Notice that F is a monotone nondecreasing function of y with $F(-\infty) = \lim_{y \to -\infty} F(y) = 0$, $F(+\infty) =$

$\lim_{y\to\infty} F(y) = 1$. The random variables $\{x_n(\omega), -\infty < n < \infty\}$ with the specification (2.1) are called independent and identically distributed in the probability literature and Bernouilli schemes in some of the discussions in ergodic theory.

Without the assumption of independence one can just say that the joint distribution function

$$P(B(n_1, y_1; \ldots; n_k, y_k)) = F_{n_1,\ldots,n_k}(y_1, \ldots, y_k)$$

is given as a function of k variables. $F_{n_1,\ldots,n_k}(y_1, \ldots, y_k)$ has the limit zero as any variable $y_i \to -\infty$ and the limit one as all y's tend to $+\infty$. Further the kth order mixed differences

$$\begin{aligned}\Delta_{h_1} \cdots \Delta_{h_k} F_{n_1,\ldots,n_k}(y_1, \ldots, y_k) \\ &= P[\{\omega : y_1 < x_{n_1}(\omega) \leqslant y_1 + h_1, \ldots, y_k < x_{n_k}(\omega) \leqslant y_k + h_k\}] \\ &\geqslant 0,\end{aligned}$$

for $h_1, \ldots, h_k \geqslant 0$. Actually the probabilities of the events given on the right are defined to be the kth order differences on the left.

A *Borel field* of sets (or events) is a collection of sets closed under complementation and countable union. The *Borel field generated by a collection of sets* is the smallest Borel field containing the collection. The reason for introducing Borel fields is that one is interested in allowing for repeated application of union, intersection and complementation to countable collections of events. Let us call the Borel field generated by all the events $B(n_1, y_1; \ldots; n_k, y_k)$, for all $k \geqslant 1$ and all y's, the Borel field $\mathscr{B}$. A nonnegative set function $P(B)$ defined on the sets $B \in \mathscr{B}$ that is completely additive for countable disjoint unions of sets

$$P\left(\bigcup_{i=1}^{\infty} B_i\right) = \sum_{i=1}^{\infty} P(B_i), \tag{2.2}$$

$B_i \in \mathscr{B}$, $B_i \cap B_j$ empty if $i \neq j$, and such that $P(\Omega) = 1$ where Ω is the set of all ω, is called a *probability measure*. A result called the Kolmogorov extension theorem states that a consistent specification of the joint distribution functions (2.2) for all finite collections of random variables uniquely determines a probability measure P on $\mathscr{B}$.

The sequence of random variables $\{x_n(\omega), -\infty < n < \infty\}$ with the governing probability measure are an example of a *stochastic* or *random process.*

It is natural to call $\mathscr{B}$ the Borel field generated by all the random variables $\{x_n(\omega), -\infty < n < \infty\}$, that is $\mathscr{B}\{x_n(\omega), -\infty < n < \infty\}$. It will also be convenient to introduce the Borel field $\mathscr{B}_n = \mathscr{B}(x_m(\omega), m \leqslant n)$ determined by conditions up to time n and the Borel field $\mathscr{F}_n = \mathscr{B}(x_m(\omega), m \geqslant n)$ determined by conditions from time n on. These Borel fields will be of interest in our discussion later on.

We now introduce the "*shift*" *transformation* τ mapping a doubly infinite vector ω onto a doubly infinite vector $\tau\omega$. τ is determined by

$$\{\tau\omega\}_n = x_n(\tau\omega) = x_{n+1}(\omega),$$

and corresponds to a shift of one unit in time. Notice that τ is an invertible transformation. τ can also be applied to sets $B \in \mathscr{B}$ by defining

$$\tau B = \{\omega : \tau^{-1}\omega \in B\}.$$

The shift transformation takes sets $B \in \mathscr{B}$ into sets $\tau B \in \mathscr{B}$. For simplicity we shall consider joint distributions $F_{n_1, \ldots, n_k}$ with property that

$$F_{n_1+1, \ldots, n_k+1}(y_1, \ldots, y_k) = F_{n_1, \ldots, n_k}(y_1, \ldots, y_k),$$

for all $k \geqslant 1$ and all y's. This is an assumption of homogeneity with respect to time, and it implies that the corresponding probability measure (assuming consistency of the family of joint distributions) P satisfies

$$P(\tau B) = P(B),$$

for all $B \in \mathscr{B}$. Such a *random process* is called a *stationary process.* In our initial discussion of a sequence of independent, identically distributed random variables the assumption of identical distributions for the coordinates implies that the corresponding process is stationary. Of course, this assumption need not have been made. However, in our development the assumption of stationarity is natural and will generally be held to.

3. GAUSSIAN PROCESSES AND MARKOV PROCESSES

A few simple but interesting examples of random processes (or sequences) are now presented. First the notion of a Gaussian (or normal process) is given. Consider the joint distribution function $F_{n_1,\ldots,n_k}(y_1,\ldots,y_k)$ of k random variables $x_{n_1}(\omega),\ldots,x_{n_k}(\omega)$. The distribution function is Gaussian (the random variables are jointly Gaussian) if F has the form

$$F_{n_1\ldots,n_k}(y_1,\ldots,y_k) = \int_{-\infty}^{y_1}\cdots\int_{-\infty}^{y_k} (2\pi)^{-k/2}|R|^{-1/2}$$
$$\times \exp\{-\tfrac{1}{2}(z-\mu)'R^{-1}(x-\mu)\}\, dz_1\cdots dz_k,$$

with μ the vector of means

$$\mu_j = E\{x_n(\omega)\} = \int x_n(\omega)P(d\omega),$$

of the random variables $x_j(\omega)$ and R the $k\times k$ matrix of covariances $r_{u,v}$ of the random variables

$$\begin{aligned} r_{u,v} &= \operatorname{cov}(x_u(\omega), x_v(\omega)) \\ &= E((x_u(\omega)-\mu_u)(x_v(\omega)-\mu_v)) \\ &= \int (x_u(\omega)-\mu_u)(x_v(\omega)-\mu_v)P(d\omega). \end{aligned}$$

Here it is assumed that the covariance matrices R are nonsingular and so strictly positive definite. $|R|$ denotes the determinant of the matrix R. The *random process* $\{x_n(\omega), -\infty < n < \infty\}$ *is called Gaussian if the joint distribution of every finite k-tuple of random variables is jointly Gaussian.* Notice that the probability structure of a Gaussian process is completely determined by its means μ_j and its covariances $r_{u,v}$. If the process is stationary then $\mu_j \equiv \mu$ and $r_{u,v} = r_{u-v}$.

Another example is that of a Markov sequence. Assume that one is given a *transition probability function* $Q(x, A)$ defined for pairs (x, A) of real numbers x and Borel sets A of real numbers. For each x, $Q(x, A)$ is a probability measure on the sets A and for each Borel set

A a Borel function of x.* The function $Q(x, A)$ is to be interpreted as the conditional probability of making an observation at time $n + 1$ in A given that one observed x at time n

$$P[x_{n+1}(\omega) \in A \mid x_n(\omega) = x] = Q(x, A).$$

Higher order transition probabilities can be defined recursively

$$\begin{aligned} Q^{(k+1)}(x, A) &= P[x_{n+k}(\omega) \in A \mid x_n(\omega) = x] \\ &= \int Q(x, du) Q^{(k)}(u, A), \\ Q^{(1)}(x, A) &= Q(x, A). \end{aligned} \tag{3.1}$$

Suppose that the process is considered at an initial time s with initial probability measure μ

$$\mu(A) = P[x_s(\omega) \in A].$$

The joint distributions are then defined by

$$\begin{aligned} &P[x_s(\omega) \in A_0, x_{s+1}(\omega) \in A_1, \ldots, x_{s+k}(\omega) \in A_k] \\ &\quad = \int_{A_0} \mu(du_0) \int_{A_1} P(u_0, du_1) \cdots \int_{A_{k-1}} P(u_{k-2}, du_{k-1}) P(u_{k-1}, A_k). \end{aligned} \tag{3.2}$$

If μ is an *invariant probability measure* with respect to Q, that is,

$$\int \mu(du) Q(u, A) = \mu(A),$$

the process can be extended backwards in time and one will get a stationary process. A simple special case is that of a Markov chain where only integer values are taken on by the random variables with positive probability. Everything can be written out in terms of transition probabilities

$$Q_{i,j} = P[x_{n+1}(\omega) = j \mid x_n(\omega) = i] \geqslant 0, \qquad \sum_j Q_{i,j} = 1.$$

The integrations in (3.1) and (3.2) then become sums.

---

* The collection of Borel sets of real numbers is the Borel field generated by the intervals on the real line. A function $g(\omega)$ is Borel if for each real z the set $\{\omega \mid g(\omega) \leqslant z\}$ is a Borel set of real numbers.

A special case of the Markovian situation will be of some interest to us. Let μ be a probability measure on Borel sets A of real numbers. Consider a real-valued function τ that is measurable in the sense that

$$\tau^{-1}A = \{x : \tau x \in A\}$$

is Borel whenever A is. At the risk of hopefully small confusion we use the symbol τ even though it has been used earlier for the shift transformation. Assume that τ is *measure-preserving* in the sense that

$$\mu(\tau^{-1}A) = \mu(A),$$

for all Borel A. This is an example of what is sometimes called a dynamical system in the ergodic theory literature.

The following two examples of dynamical systems are of some interest:

Example 1. Here μ is Lebesgue measure (length) on the unit interval $[0, 1)$ with $\tau x = x + \theta$ modulo one.

Example 2. μ is again Lebesgue measure on $[0, 1)$ with $\tau x = 2x$ modulo one.

Various limit theorems are of especial interest in probability theory. We shall mention one of these, usually called the *central limit theorem.* Suppose that the random variables $x_j(\omega)$, $j = \ldots, -1, 0, 1, \ldots$, are independent and identically distributed with common mean $m \equiv Ex_j(\omega) = \int x_j(\omega)\, dP(\omega)$ and variance $\sigma^2 = E(x_j(\omega) - m)^2 > 0$. There are then a number of ways of showing that

$$P\left\{\sum_{j=1}^{n} (x_j - m)/\sqrt{n}\sigma \leqslant x\right\} \to \Phi(x) = \int_{-\infty}^{x} \frac{e^{-u^2/2}}{\sqrt{2\pi}}\, du,$$

as $n \to \infty$ for each real x, that is, the partial sums of the x_j's are asymptotically normally distributed. One would like to find conditions under which such a result would still hold when the x_j's are stationary but not necessarily independent. It is not enough to just assume ordinary ergodicity or mixing (see Section 4). This has in part motivated the interest in stronger versions of ergodicity and mixing. There are still open questions and interesting research on the central limit theorem for dependent processes.

4. ERGODICITY AND MIXING

Suppose that $\{x_n(\omega), -\infty < n < \infty\}$ is a stationary process with probability measure P. The process is said to be *ergodic* if for any two given events $B, B' \in \mathscr{B}$

$$\frac{1}{n}\sum_{k=1}^{n} P(B \cap \tau^k B') \to P(B)P(B'), \tag{4.1}$$

as $n \to \infty$. If the two events B, B' were independent we would have $P(B \cap B') = P(B)P(B')$. The limiting relation (4.1) says that as B' is shifted further and further into the future ($\tau^k B'$ is the kth shift of B' forward in time), one has an approach to statistical independence of B on the average. An event B is called an *invariant event* if $B = \tau B$ (except for a possible exceptional set of probability zero). Notice that ergodicity implies that the only invariant events are those of probability zero or one. For if B is invariant, by ergodicity we have taking $B' = B$ in (4.1)

$$\frac{1}{n}\sum_{k=1}^{n} P(B \cap \tau^k B) = P(B) = P(B)^2.$$

The only invariant events are then the trivial ones, the empty event of probability zero and the sure event of probability one. One can conversely say that if the only invariant sets are the trivial ones then the stationary process is ergodic. This can be shown to follow from the famous ergodic theorem which states that if $x(\omega)$ is a random variable on the probability space with finite mean

$$E\{x(\omega)\} = \int x(\omega)P(d\omega),$$

then

$$\frac{1}{n}\sum_{k=1}^{n} x(\tau^k\omega) \to \hat{x}(\omega), \tag{4.2}$$

as $n \to \infty$ with probability one and the limiting random variable $\hat{x}(\omega)$

is constant if the process is ergodic and is then equal to $E\{x(\omega)\}$ (see [4]). The notion of ergodicity can also be applied to a dynamical system. The same formulation and results hold for that context with τ the corresponding function and μ replacing P. Example 1 with θ a rational k/m (k, m positive integers) is an example of a nonergodic system. Any set containing with x all its translates $x + j/m$, $j = 0, 1, \ldots, m - 1$, modulo one is clearly invariant. There are many such nontrivial invariant sets. However, if θ is irrational, the system is ergodic. This can be seen by considering the normalized sum (4.2) for functions

$$y(x) = \exp(2\pi ikx), \qquad k = 0, \pm 1, \ldots,$$

complex exponentials since any integrable function can be approximated in mean by finite linear combinations of exponentials. Now if $k \neq 0$,

$$\frac{1}{n}\sum_{j=1}^{n} y(\tau^j x) = \frac{1}{n}\sum_{j=1}^{n} \exp(2\pi ikx)\exp(2\pi ijk\theta)$$

$$= \exp(2\pi ikx)\,\frac{1}{n}\,\frac{e^{2\pi ikn\theta} - 1}{e^{2\pi ik\theta} - 1}\,e^{2\pi ik\theta} \to 0 = \int_0^1 \exp(2\pi ikx)\,dx,$$

as $n \to \infty$ while if $k = 0$ the normalized partial sum is

$$1 = \int_0^1 dx.$$

A stationary process is said to be *mixing* if for any two events $B, B' \in \mathscr{B}$

$$P(B \cap \tau^k B') \to P(B)P(B'), \tag{4.3}$$

as $n \to \infty$. This is stronger than ergodicity since the limiting relationship is assumed to hold without averaging. As before, the concept of mixing can also be applied to a dynamical system. Example 1 with θ irrational is ergodic but not mixing.

Let us consider the case of a stationary sequence of independent and identically distributed random variables. That this process is mixing can be seen from the following argument. It is clear that if B

and B' are events determined by conditions on a finite number of random variables, then

$$P(B \cap \tau^k B') = P(B)P(B'),$$

if τ is sufficiently large. Then (4.3) can be shown to be valid for any two events by an approximation argument. One can show that given any event B and number $\epsilon > 0$, there is an integer $n = n(\epsilon)$ and a corresponding event B_n determined by conditions only on $x_j(\omega)$, $|j| \leqslant n(\epsilon)$, such that

$$P(B \ominus B_n) < \epsilon,$$

where $B \ominus B_n = (B - B_n) \cup (B_n - B)$, the symmetric difference of B and B_n. Notice that if $\{x_n(\omega), -\infty < n < \infty\}$ is ergodic or mixing, any derived process $\{y_n(\omega), -\infty < n < \infty\}$ derived by a reasonable linear or nonlinear function (a measurable function*)

$$y_n(\omega) = f(\tau^n \omega), \qquad n = \ldots, -1, 0, 1, \ldots,$$

is also correspondingly ergodic or mixing.

It's of some interest to look at Gaussian stationary processes in this context. Assume that a Gaussian stationary process $\{x_n(\omega)\}$ has mean $\mu \equiv 0$ and covariance sequence r_n, $n = 0, \pm 1, \ldots$. The covariance sequence is positive definite in the sense that

$$\sum_{j,k=1}^{m} c_j r_{j-k} \bar{c}_k = E\left|\sum_{j=1}^{m} c_j x_j\right|^2 \geqslant 0,$$

for any complex c's and any integer $m \geqslant 1$. By an old result of Herglotz there is a bounded nondecreasing function G such that

$$r_n = \int_{-\pi}^{\pi} e^{in\lambda}\, dG(\lambda).$$

G is called the spectral distribution function of the process and if it is differentiable, its derivative $g(\lambda) = G'(\lambda)$ is called the spectral density of the process. Maruyama [10] and Grenander have shown that a *Gaussian stationary process is ergodic if and only if its spectral*

---

* f is measurable if $f^{-1}(A) \in \mathscr{B}$ for any Borel set A of real numbers.

distribution function G is a continuous function of λ (i.e., has no jumps). For if G has a jump at μ, one can show that

$$\frac{1}{n}\sum_{k=1}^{n} x_k^2(\omega)$$

has a limiting random variable $y(\omega)$ in mean square as $n \to \infty$. $y(\omega)$ is an *invariant* random variable ($y(\tau\omega) = y(\omega)$) with positive variance and so any set of the form $\{\omega : y(\omega) < z\}$ with $z > 0$ is a nontrivial invariant set. However, if G is continuous one can show that for any polynomial $p(\omega)$ in the random variables $\{x_n(\omega)\}$

$$\frac{1}{n}\sum_{j=1}^{n} p(\tau^j\omega) \to E\{p(\omega)\}$$

in mean square as $n \to \infty$. Since every random variable with finite mean can be approximated by such a polynomial, there are no nontrivial invariant random variables. Hence there are no nontrivial invariant sets.

Again, by an approximation argument one can show that *a stationary Gaussian process is mixing if and only if the covariance sequence*

$$r_n = E\{x_0(\omega)x_n(\omega)\} \to 0,$$

as $n \to \infty$.

5. STRONG MIXING AND UNIFORM ERGODICITY

The concepts of asymptotic independence embodied in the notions of ergodicity and mixing as already noted are not strong enough to obtain limit theorems like the central limit theorem for dependent random variables. For this and other reasons stronger conditions have been introduced and examined at some length.

A concept that I called strong mixing was introduced in [12]. Consider a stationary process with the corresponding Borel fields $\mathscr{B}_n$ and $\mathscr{F}_n$. The process is strongly mixing if

$$\sup_{\substack{B\in\mathscr{B}_n \\ F\in\mathscr{F}_{n+k}}} |P(B \cap F) - P(B)P(F)| = \alpha(k) \downarrow 0, \tag{5.1}$$

as $k \to \infty$. The coefficient $\alpha(k)$ is a measure of how close the Borel fields $\mathscr{B}_n$ and $\mathscr{F}_{n+k}$ are to independence. In their investigation of strong mixing for stationary Gaussian sequences Kolmogorov and Rozanov [7] noted that *in the Gaussian case condition* (5.1) *is equivalent to*

$$\sup_{\substack{y \in L^2(\mathscr{B}_n) \\ z \in L^2(\mathscr{F}_{n+k})}} |\mathrm{corr}(y(\omega), z(\omega))| = \beta(k) \downarrow 0, \tag{5.2}$$

as $k \to \infty$ where

$$\mathrm{corr}(y(\omega), z(\omega)) = \frac{E(y(\omega)z(\omega)) - Ey(\omega)Ez(\omega)}{\{E(y(\omega) - Ey(\omega))^2 E(z(\omega) - Ez(\omega))^2\}^{1/2}}$$

is the correlation coefficient of y and z. $L^2(\mathscr{B}_n)$ and $L^2(\mathscr{F}_{n+k})$ are the families of square integrable random variables measurable with respect to the Borel fields $\mathscr{B}_n$ and $\mathscr{F}_{n+k}$ respectively. The elements of $L^2(\mathscr{B}_n)$ and $L^2(\mathscr{F}_{n+k})$ are functions of $\{x_j(\omega), j \leqslant n\}$ and $\{x_j(\omega), j \geqslant n + k\}$. Condition (5.2) states that the spaces $L^2(\mathscr{B}_n)$ and $L^2(\mathscr{F}_{n+k})$ become uniformly orthogonal as $k \to \infty$. This condition in the non-Gaussian case is more stringent than strong mixing. Suppose the spaces $L^2(\mathscr{B}_n)$ and $L^2(\mathscr{F}_{n+k})$ are replaced by the subspaces $\mathscr{M}^n_{-\infty} = \mathscr{M}\{x_j(\omega), j \leqslant n\}$, $\mathscr{M}^\infty_{n+k} = \mathscr{M}\{x_j(\omega), j \geqslant n + k\}$ consisting of limits in mean square of finite linear forms in $\{x_j(\omega), j \leqslant n\}$ and $\{x_j(\omega), j \geqslant n + k\}$ respectively. The spaces $\mathscr{M}^n_{-\infty}$ are smaller than the spaces $L^2(\mathscr{B}_n)$ introduced earlier. In the Gaussian stationary case

$$\sup_{\substack{y \in \mathscr{M}^n_{-\infty} \\ z \in \mathscr{M}^\infty_{n+k}}} |\mathrm{corr}\,(y(\omega), z(\omega))| = \beta(k),$$

so that the maximal correlation is obtained just by considering linear forms in the Gaussian random variables. Kolmogorov and Rozanov use this remark to show that *a Gaussian stationary process with a differentiable spectral distribution function and a continuous spectral density bounded away from zero is strongly mixing.* Helson and Sarason later used it to obtain a necessary and sufficient condition for a stationary Gaussian sequence to be strongly mixing. We shall discuss their condition later on.

It is clear that if a stationary process is strongly mixing then it is mixing. The converse doesn't hold as one can see by looking at the

Markov process suggested by example 2. Let μ be Lebesgue measure on $[0, 1)$ with the transition function determined by the function $\tau x = 2x$ modulo one, that is, $x_{n+1}(\omega) = 2x_n(\omega)$ modulo one. This determines a stationary Markov process which is mixing. However, it is certainly not strongly mixing since it evolves deterministically.

It is also of some interest to see what strong mixing means for a stationary Markov process. Let μ be an invariant probability measure for the transition probability $Q(\cdot, \cdot)$. μ and Q determine the structure of a stationary Markov process. Introduce the operator T on bounded measurable functions h determined by Q

$$(Th)(x) = \int Q(x, dy)h(y).$$

T can be extended to an operator acting on functions integrable in pth ($p \geqslant 1$) mean with respect to μ. The L^p norm of such a function h is

$$\|h\|_p = \left\{\int |h(x)|^p \mu(dx)\right\}^{1/p},$$

if $\infty > p \geqslant 1$ and is the essential sup

$$\|h\|_\infty = \text{ess sup } |h| = \lim_{p\to\infty} \|h\|_p,$$

if $p = \infty$. Notice that

$$\left|\int Q(x, dy)h(y)\right| \leqslant \left\{\int Q(x, dy)|h(y)|^p\right\}^{1/p}.$$

This together with the fact that μ is invariant relative to Q implies that T is a contraction operator for L^p, $\infty \geqslant p \geqslant 1$,

$$\|Th\|_p \leqslant \|h\|_p.$$

One can show (see [13]) that *the stationary Markov process determined by μ and Q is strongly mixing if and only if*

$$\sup_{h\perp 1} \frac{\|T^n h\|_1}{\|h\|_\infty} \to 0, \tag{5.3}$$

as $n \to \infty$, where by $h \perp 1$ we mean $\int h(x)\mu(dx) = 0$. In the case of a stationary Markov chain (countable state) strong mixing and mixing *are identical.* We have already seen an example of stationary Markov processes with a more complicated state space where this is not the case. Another interesting example of this type is now given. Consider the following model of random rotations. Points on the circle are represented by points $x \in [0, 1)$ and rotation by y units as $x + y$ modulo one. If $x_n(\omega)$ is given, the potential observation at time $n + 1$ is given by

$$x_{n+1}(\omega) = x_n(\omega) + y_n(\omega),$$

where $y_n(\omega)$ is statistically independent of $x_n(\omega)$ and is assumed to have probability distribution η. This corresponds to a Markov process with transition probability

$$Q(x, A) = \eta(A - x).$$

One can see that the uniform distribution (Lebesgue measure) is an invariant measure relative to Q and suppose we take it as μ. Suppose we let $h(x) = e^{2\pi ikx}$, $k \neq 0$. The complex exponentials are bounded and orthogonal to one. Notice that

$$\begin{aligned}(T^n h)(x) &= \int_0^1 Q^{(n)}(x, dy)\, e^{2\pi iky}\, dy \\ &= \int_0^1 \eta^{(n)}(dy - x)\, e^{2\pi iky}\, dy \\ &= e^{2\pi ikx}\hat{\eta}_k,\end{aligned}$$

where $\eta^{(n)}$ is the nth convolution of η with itself and

$$\hat{\eta}_k = \int_0^1 e^{2\pi iky}\eta(dy).$$

Condition (5.3) implies that

$$\sup_{k \neq 0} |\hat{\eta}_k| < 1 \tag{5.4}$$

is a necessary condition for the Markov process to be strongly mixing. One can show that (5.4) is not only necessary but also sufficient for the process to be strongly mixing. However, a necessary and sufficient condition for the process to be mixing is that the probability measure η not be such that all its mass is located on points of the form

$$\theta + \frac{k}{m} \text{ modulo one}$$

$k = 0, 1, \ldots, m - 1$ for some fixed positive integer m and real number θ. It is clear that for this set of Markov processes strong mixing is in fact much stronger than mixing. Also, the condition of uniform asymptotic orthogonality of past and future (5.2) can be shown to be equivalent to

$$\sup_{h \perp 1} \frac{\|T^n h\|_2}{\|h\|_2} < 1, \tag{5.5}$$

for some integer $n \geqslant 1$ in the case of a stationary Markov process. This is appreciably stronger than strong mixing (5.1) in the Markov case. The following Markov chain is mixing (hence strongly mixing) but does not satisfy (5.5). The states or values of the chain are indicated by double indexing rather than integers alone. Let the states be 0 and (j, k) with $j = 1, 2, \ldots,$ and $k = 0, 1, \ldots, j - 1$. Let the transition probabilities be

$$Q_{0,(j,0)} = q_j > 0, \sum q_j = 1$$

$$Q_{(j,k),(j,k+1)} = 1 \qquad \text{for } k = 0, \ldots, j-2$$

$$Q_{(j,j-1),0} = 1$$

with all other transition probabilities set equal to zero. Assume that $\sum jq_j < \infty$. The stationary probability distribution is

$$\mu_0 = (1 + \sum jq_j)^{-1}$$

$$\mu_{(j,0)} = q_j \mu_0$$

$$\mu_{(j,k)} = \mu_{(j,k+1)} \qquad k = 0, 1, \ldots, j-2.$$

Let ${}_jh$ be the function with ${}_jh_s = 0$ unless $s = (j, j-1), (j, j-2)$ where

$${}_jh_{(j,j-1)} = (2\mu_0 q_j)^{-1/2} = -{}_jh_{(j,j-2)}.$$

Since

$$\|{}_jh\|_2 = 1, \qquad \|T^{j-2}{}_jh\|_2 = 1,$$

it follows that

$$\sup_{h \perp 1} \frac{\|T^n h\|_2}{\|h\|_2} = 1,$$

for all n. It is clear that if a stationary Markov process satisfies condition (5.5) then the sequence $\beta(k)$ decreases to zero exponentially fast as $k \to \infty$.

Another mixing condition has been introduced by Ibragimov [5]:

$$|P(B \cap F) - P(B)P(F)| \leqslant \gamma(k)P(B), \tag{5.6}$$

for all $B \in \mathscr{B}_n$, $F \in \mathscr{F}_{n+k}$ with $\gamma(k) \downarrow 0$ as $k \to \infty$. This condition is obviously much more stringent than strong mixing. It is even stronger than the uniform asymptotic orthogonality of past and future; and so in the case of a stationary Markov process satisfying (5.6), the sequence $\gamma(k)$ tends to zero exponentially as $k \to \infty$. It is an interesting question as to whether this exponential decrease holds for any stationary process satisfying (5.6).* One can easily show that a necessary and sufficient condition for a Gaussian process to satisfy Ibragimov's condition is that it be finite step dependent. A process is finite step dependent if there is some integer $s \geqslant 1$ such that for any $B \in \mathscr{B}_n$ and any $F \in \mathscr{F}_{n+s}$, B and F are independent

$$P(B \cap F) = P(B)P(F).$$

A strengthening of ergodicity was introduced by Cogburn [2] in a discussion of limit theorems for partial sums of Markov sequences. He calls a stationary process uniformly ergodic if

$$\sup_{\substack{B \in \mathscr{B}_0 \\ F \in \mathscr{F}_0}} \left| \frac{1}{n} \sum_{k=1}^{n} P(B \cap \tau^k F) - P(B)P(F) \right| = \bar{\alpha}(k) \downarrow 0,$$

---

* Recently H. Kesten and G. L. O'Brien (see *Duke Math. J.*, **43** (1976) 405–415) and still later R. C. Bradley, Jr., independently have shown that given any sequence $\varphi(k) \to 0$ (however slowly) as $k \to \infty$, there is a stationary non-Markovian process that satisfies (5.6) and such that $\gamma(k) \geqslant \varphi(k)$.

as $k \to \infty$. This represents a strengthening of ergodicity parallel to the way in which strong mixing strengthens mixing. A process has a cyclically moving sequence of sets with cycle length $l > 1$ if there is an event C such that

$$C, \tau C, \ldots, \tau^{l-1}C,$$

form a disjoint partition of the probability space and $\tau^l C = C$. The following interesting result was obtained in [14]. *A uniformly ergodic stationary process with no cyclically moving sequence of sets is strongly mixing.* This is a rather surprising result because it states that there isn't much difference between uniform ergodicity and strong mixing. This is in sharp contrast to the difference between ergodicity and mixing which is quite broad.

6. MORE ON GAUSSIAN PROCESSES

We return to discussion of strong mixing for Gaussian stationary processes and give an interesting necessary and sufficient condition obtained by Helson and Sarason [3, 15]. They show that the spectral distribution function G of the Gaussian process must be differentiable with the spectral density $g(\lambda) = G'(\lambda)$ having the form

$$g(\lambda) = |P(e^{i\lambda})|^2 \omega(\lambda),$$

with

$$P(e^{i\lambda}) = \sum_{|j| \leq n} c_j e^{ij\lambda}$$

a finite trigonometric polynomial and

$$\omega(\lambda) = \exp\{u(\lambda) + \tilde{v}(\lambda)\}, \tag{6.1}$$

where u, v are continuous and $\tilde{v}$ the conjugate function of v. $\tilde{v}$ is the conjugate function of v if it is given by the integral formula

$$\tilde{v}(\lambda) = -\frac{1}{\pi} \int_0^{\pi} \frac{v(\lambda + t) - v(\lambda - t)}{2 \tan \frac{1}{2}t} \, dt,$$

where v is understood to be defined by periodicity (with period 2π) if its argument falls out of the range $[-\pi, \pi)$. Actually Sarason has shown that a concept of some interest in recent work in harmonic analysis is of some relevance. Given an integrable function h and a finite interval I let

$$h_I = |I|^{-1} \int_I h(x)\, dx,$$

where $|I|$ is the length of I. Set

$$M_a(h) = \sup_{|I| \leq a} |I|^{-1} \int_I |h(x) - h_I|\, dx,$$

and

$$M_0(h) = \lim_{a \to 0} M_a(h).$$

The function h is said to have *vanishing mean oscillation* if $M_0(h) = 0$. One of the results in [16] indicates that nonnegative ω has the form (6.1) with u, v continuous if and only if $\log \omega$ is of vanishing mean oscillation (VMO). An alternative characterization can be given in the following manner. Let

$$N_a(\omega) = \sup_{|I| \leq a} |I|^{-2} \left(\int_I \omega(x) dx \right) \left(\int_I \omega(x)^{-1}\, dx \right),$$

and set

$$N_0(\omega) = \lim_{a \to 0} N_a(\omega).$$

Then one can show that $\log \omega$ is VMO if and only if $N_0(\omega) = 1$. Examples of strong mixing Gaussian processes with discontinuous spectral densities can be constructed using these conditions.

It would clearly be of some interest to determine necessary and sufficient conditions for strong mixing for a continuous time parameter Gaussian stationary process and also for one with a multidimensional time parameter. A discussion of some earlier work on strong mixing for Gaussian processes can be found in a paper of Yaglom [17].

7. THE CENTRAL LIMIT THEOREM

Many people have worked on conditions under which the central limit theorem is still valid for a stationary process. We shall just mention some simple conditions (see [5], [9]). Let $\{X_j\}$ be a stationary process with means $EX_j \equiv 0$. Assume that the βth moment

$$E|X_j|^\beta < \infty,$$

for some β, $2 \leqslant \beta < \infty$. Further, let the process be mixing in the sense of Ibragimov with

$$\sum_{k=1}^{\infty} \gamma(k)^{1-1/\beta}, \tag{7.1}$$

or strongly mixing with $\beta > 2$ and

$$\sum_{k=1}^{\infty} \alpha(k)^{1-2/\beta} < \infty. \tag{7.2}$$

One can then show that

$$\frac{1}{n} E\left(\sum_{j=1}^{n} X_j\right) \rightarrow \sigma^2 \geqslant 0,$$

as $n \rightarrow \infty$ and if $\sigma^2 > 0$ that

$$\frac{1}{\sqrt{n}\sigma} \sum_{j=1}^{n} X_j,$$

is asymptotically normally distributed with mean zero and variance one. Actually one can show much more in this context. Let

$$S_k = \sum_{j=1}^{n} X_j.$$

Consider functionals of the partial sums S_k like

$$\max_{k=1,\ldots,n} \left|\frac{S_k}{n^{1/2}\sigma}\right|,$$

or

$$\sum_{k=1}^{n} \left(\frac{S_k}{n^{1/2}\sigma}\right)^2 \frac{1}{n}.$$

There are a number of applications in which the asymptotic distributions of such functionals as $n \to \infty$ is of interest. We can give a general formulation which includes these specific functionals. Introduce

$$S_{[nt]}, \qquad 0 \leqslant t \leqslant 1,$$

where $[u]$ is the greatest integer less than or equal to u. Notice that $S_{[nt]}$ is a continuous parameter random process. Let ϕ be a functional that is defined for functions that are continuous except for a finite number of possible jumps. Assume that ϕ is continuous with respect to the supremum norm, that is, given $\epsilon > 0$ there is a $\delta(\epsilon) > 0$ such that if

$$\sup |x(\cdot) - y(\cdot)| < \delta,$$

then

$$|\phi(x(\cdot)) - \phi(y(\cdot))| < \epsilon.$$

Our interest is then in the asymptotic distribution of

$$\phi\left(\frac{S_{[nt]}}{n^{1/2}\sigma}; \qquad 0 \leqslant t \leqslant 1\right), \tag{7.3}$$

as $n \to \infty$. In order to formulate the desired limit theorem, the Brownian motion (or Wiener) process must be introduced. The Brownian motion process $B(t)$, $0 \leqslant t$, is a Gaussian process (all joint distributions at a finite number of time points are Gaussian) with mean

$$EB(t) \equiv 0,$$

and covariance function

$$E[B(t)B(\tau)] = \min(t, \tau),$$

$0 \leqslant t, \tau$. The probability measure of a Brownian motion process $B(t)$ can be thought of as a measure on the space of continuous functions. Under the conditions (7.1) or (7.2) one can show (see [1] and [9]) that

$$P\left\{\phi\left(\frac{S_{[nt]}}{n^{1/2}\sigma}; 0 \leqslant t \leqslant 1\right) \leqslant x\right\} \to P\{\phi(B(t); 0 \leqslant t \leqslant 1) \leqslant x\},$$

as $n \to \infty$. Thus the probability distribution of the functional (7.3) tends to that of the corresponding functional of the Brownian motion process as $n \to \infty$. The process $B(t)$ is a Markov process and the distribution of many functionals of $B(\cdot)$ can be explicitly computed [6].

Various related notions of mixing are of use in obtaining other limit theorems [8] and even in obtaining interesting results on solutions of stochastic differential equations [11] that arise in a number of fields of application.

REFERENCES

1. P. Billingsley, *Convergence of Probability Measures*, Wiley, New York, 1968.
2. R. Cogburn, "Conditional probability operators," *Ann. Math. Stat.*, **33** (1962), 634–658.
3. H. Helson and D. Sarason, "Past and future," *Math. Scand.*, **30** (1967), 5–16.
4. P. R. Halmos, *Lectures on Ergodic Theory*, Tokyo, 1956.
5. I. A. Ibragimov and Yu. V. Linnik, *Independent and Stationary Sequences of Random Variables*, North Holland, Gröningen, 1971.
6. M. Kac, "On some connections between probability theory and differential and integral equations," *Proc. 2nd Berkeley Symposium Math. Stat. and Prob.* (1951), 189–215.
7. A. N. Kolmogorov and Yu. A. Rozanov, "On a strong mixing condition for stationary Gaussian processes," *Theor. Probability Appl.*, **5** (1960), 204–208.
8. M. R. Leadbetter, "Extreme value theory under weak mixing conditions," this study.
9. D. L. McLeish, "Invariance principles for dependent variables," *Z. Wahrscheinlichkeitstheorie verw. Geb.*, **32** (1975), 165–178.
10. G. Maruyama, "The harmonic analysis of stationary stochastic processes," *Mem. Fac. Sci, Kyushu Univ.*, **A4** (1949), 45–106.
11. G. Papanicolaou, "Asymptotic analysis of stochastic equations," this study.
12. M. Rosenblatt, "A central limit theorem and a strong mixing condition," *Proc. Nat. Acad. Sci. U.S.A.*, **42** (1956), 43–47.
13. ———, *Markov Processes. Structure and Asymptotic Behavior*, Springer-Verlag, Berlin, 1971.

14. ———, "Uniform ergodicity and strong mixing," *Z. Wahrscheinlichkeitstheorie verw. Geb.*, **24** (1972), 79–84.

15. D. Sarason, "An addendum to 'Past and future'," *Math. Scand.*, **30** (1972), 62–64.

16. ———, "Functions of vanishing mean oscillation," *Trans. Amer. Math. Soc.*, **207** (1975), 391–405.

17. A. M. Yaglom, "Stationary Gaussian processes satisfying the strong mixing condition and best predictable functionals," *Proc. Internat. Res. Sem. Statist. Lab.*, University of California, Berkeley, 1963, Springer-Verlag, New York, 1965, 241–252.

EXTREME VALUE THEORY UNDER WEAK MIXING CONDITIONS

*M. R. Leadbetter**

1. SELECTED TOPICS FROM CLASSICAL EXTREME VALUE THEORY

1.1. Introduction. The practical importance of the statistical properties of *extreme values* is clear and has long been recognized (cf. [18]). The maximum daily rainfall in a given period has obvious agricultural implications. Equipment must be designed to work under maximum likely stresses. Dams need to be built to cope with maximal water flow, and so on.

Classical extreme value theory represents the quantities of interest as a sequence $\xi_1, \xi_2, \ldots,$ of independent and identically distributed (i.i.d.) random variables (r.v.'s). Substantial attention in this theory is focussed on the distribution of $M_n = \max(\xi_1, \xi_2, \ldots, \xi_n)$,

* Research noted supported in part by the Office of Naval Research under Contract N00014-75C-0809.

particularly as $n \to \infty$. (The relation $\min(\xi_1, \xi_2, \ldots, \xi_n) = -\max(-\xi_1, -\xi_2, \ldots, -\xi_n)$ then gives corresponding results for minima.) In this paper we shall be concerned with conditions under which the theory generalizes to apply to the maxima (and related quantities) of both discrete and continuous parameter stochastic processes. Specifically we shall, in the discrete case, consider $M_n = \max(\xi_1, \ldots, \xi_n)$ as above, but where now the r.v.'s ξ_i are no longer assumed independent. That is, $\{\xi_n : n \geqslant 1\}$ is a discrete parameter stochastic process. In the continuous parameter case we shall consider $M(T) = \sup\{\xi(t) : 0 \leqslant t \leqslant T\}$ where $\{\xi(t) : t \geqslant 0\}$ is a continuous parameter process. In both cases we shall be concerned with finding *dependence conditions* under which the classical theory involving the distribution of M_n or $M(T)$ still substantially applies (for large n or T). Primarily for simplicity, we shall still assume that the r.v.'s considered are identically distributed (indeed that the processes are stationary), though this assumption could also be weakened.

Returning, then, to the classical case, let $\{\xi_n : n = 1, 2, \ldots\}$ be i.i.d. random variables with the common distribution function (d.f.) F and write $M_n = \max\{\xi_1, \xi_2, \ldots, \xi_n\}$. In this case since (for any fixed x), the event $\{M_n \leqslant x\}$ is precisely the same as $\{\xi_1 \leqslant x, \xi_2 \leqslant x, \ldots, \xi_n \leqslant x\}$, the d.f. of M_n is given by

$$P\{M_n \leqslant x\} = P\{\xi_1 \leqslant x, \xi_2 \leqslant x, \ldots, \xi_n \leqslant x\} \tag{1.1.1}$$

$$= F^n(x). \tag{1.1.2}$$

In principle the distribution of M_n is thus completely specified for any given n, though of course considerable computational effort may be required in application, and each new F will give rise to a new form for F^n. However it has been found that, when n is large, the distribution of M_n can (in a sense to be made precise below) behave in only a limited number (essentially three, in fact) distinct ways. In this chapter we shall explicitly explore this and other related topics which are central to the classical theory. For example we shall also look at the corresponding results for the kth largest $M_n^{(k)}$ of $\xi_1, \ldots, \xi_n$, both when k is fixed and when $k \to \infty$ with n.

There are many other aspects to classical extreme value theory (e.g., involving almost sure convergence, stability, and so on), but the distributional properties of M_n (and $M_n^{(k)}$) are central and will suffice for our purposes in illustrating the theory and its extensions.

The main purpose of this paper is to explore ways in which the independence assumption of the classical theory may be weakened, since in applications the r.v.'s of interest are very often dependent, even in "discrete time" (and certainly for continuous time). We shall still, as noted, assume that the r.v.'s are identically distributed—and indeed stationary (though it will be clear that the theory applies to more general cases under sufficient assumptions).

In Chapter 2 we shall discuss conditions under which the classical theory still substantially applies, under certain dependence restrictions, for (stationary) *sequences* of r.v.'s. The restrictions on dependence will be a very weak (distributional) form of *mixing* condition (cf. [34] in this volume for a general discussion of such conditions). Roughly they ensure that the "dependence" between the "past" ξ_j's before $j = m$ and the "future" ξ_j's after $j = n$ falls off appropriately quickly as their "separation" $n - m$ increases. It will be shown how stationary *normal* sequences fit into this theory and how the appropriate mixing condition is satisfied under very weak assumptions concerning the covariances between the r.v.'s—a natural way of describing dependence for normal processes.

A quantity closely related to $M_n = \max(\xi_1, \xi_2, \ldots, \xi_n)$ is the number S_n $(=S_n(u))$ of *exceedances* of the level u by $\xi_1, \ldots, \xi_n$, i.e., the number of i $(1 \leqslant i \leqslant n)$ such that $\xi_i > u$. Clearly $P\{M_n \leqslant u\} = P\{S_n = 0\}$ or, more generally $P\{M_n^{(k)} \leqslant u\} = P\{S_n \leqslant k - 1\}$ $(k = 1, 2, \ldots, n)$. These relations will be discussed and used in both Chapters 1 and 2 where it will be shown that, under general conditions, S_n has a Poisson limit as $u \to \infty$ in a way which is coordinated with n. This theme will be developed further in Chapter 2 where the exceedances of a level will be regarded as a *point process*. As will be indicated there, this point process is approximately Poisson if the level u is high—a fact which implies further results concerning the $M_n^{(k)}$ and related quantities.

Chapters 1 and 2 are concerned with *sequences* of r.v.'s $\{\xi_n : n = 1, 2, \ldots\}$. In Chapter 3 we shall turn to corresponding problems in "continuous time"—e.g., concerning the distribution of

$M(T) = \sup\{\xi(t) : 0 \leqslant t \leqslant T\}$ where $\xi(t)$ is a r.v. for each t (i.e., $\{\xi(t) : t \geqslant 0\}$ is a *stochastic* process). Generally there is no hope of evaluating the distribution of $M(T)$ explicitly for finite values of T (even for finite T we are dealing with the supremum of *infinitely* many (nonindependent) r.v.'s $\xi(t)$). However again we find—at least under a mixing condition—that the maximum $M(T)$ exhibits just the same three types of asymptotic behavior as for the classical case of i.i.d. sequences.

We shall deal with such questions in Chapter 3 along with other general properties of $M(T)$. The properties of $M(T)$ are known in most detail when $\{\xi(t)\}$ is a stationary *normal* process. This case will also be considered in Chapter 3, where the known results concerning $M(T)$ and related quantities are indicated.

Though we do not consider them here, some results are also known when t is a multidimensional parameter (cf. [45, 46]). Also, it should be noted that, apart from the natural and obvious interest in $M(T)$ in cases where $\xi(t)$ models some physical process, applications of the asymptotic normal theory occur in other statistical contexts. For example, interesting use of the theory is made by D. G. Kendall [49] in estimation problems in archaeological studies, and by Bickel and Rosenblatt (cf. [47]) in connection with probability density estimation.

Finally we note that our attempt, in this paper, is to provide an account of the essential features of this branch of extreme value theory and the use of mixing conditions, but without too many details. For an extended and more detailed review of these and other related topics, the reader is referred to [23].

1.2. Gnedenko's Theorem and related topics. As noted in the previous section, classical extreme value theory is concerned with i.i.d. sequences $\{\xi_1, \xi_2, \dots\}$, and a central problem is that of determining the distribution of $M_n = \max(\xi_1, \xi_2, \dots, \xi_n)$, especially when n becomes large. Of course in general M_n will become larger as n increases, and will not itself have a limiting distribution (i.e., $M_n \to x_0$ almost surely (a.s.) where x_0 ($\leqslant \infty$) is the right-hand endpoint of the common d.f. F, as is easy to prove). However, often M_n may be normalized by appropriate sequences $\{a_n\}$, $\{b_n\}$ so that $a_n(M_n - b_n)$ does have a nondegenerate limiting distribution. The following theorem of Gnedenko [16] is central to the subject and shows that there are only three possible forms of such limiting distributions.

THEOREM 1.2.1 (Gnedenko): *Let* $\{\xi_n, n = 1, 2, \ldots\}$ *be a sequence of* i.i.d. *random variables and write* $M_n = \max(\xi_1, \xi_2, \ldots, \xi_n\}$. *Suppose that for some sequences* $\{a_n > 0\}$, $\{b_n\}$ *of real constants,* $a_n(M_n - b_n)$ *converges in distribution to a* r.v. *with nondegenerate* d.f. G. *Then the* d.f. G *must be of one of the following three "types":*

$$\begin{array}{rlll} \textit{Type I:} & G(x) = \exp(-e^{-x}) & & -\infty < x < \infty \\ \textit{Type II:} & G(x) = 0 & & x \leqslant 0 \\ & \quad = \exp(-x^{-\alpha}) & \textit{for some } \alpha > 0 & x > 0 \\ \textit{Type III:} & G(x) = \exp(-(-x)^{\alpha}) & \textit{for some } \alpha > 0 & x \leqslant 0 \\ & \quad = 1 & & x > 0. \end{array}$$

(In this statement, by a "type" we mean all d.f.'s which can be obtained from those indicated by replacing x by $ax + b$ for any $a > 0, b$, e.g., Type I includes all d.f.'s of the form $\exp\{-e^{-(ax+b)}\}$ for any fixed $a > 0, b$.)

Detailed proofs of this theorem may be found in [16], [19]. We do not give these details here, but it will be useful to outline the main ideas of the proof in the following three lemmas, the first of which is due to Khintchine (and its proof may be found in [17, Section 10, Theorem 1]).

LEMMA 1.2.2: *If a sequence* $\{G_n\}$ *of* d.f.'s *converges as* $n \to \infty$ *to a nondegenerate* d.f. G, *and if* $\{\lambda_n > 0\}$, $\{\mu_n\}$ *are real constants, then the sequence* $G_n(\lambda_n x + \mu_n)$ *can converge to a nondegenerate* d.f. *only if this is of the same "type" as* G, *i.e., is* $G(\lambda x + \mu)$ *for some* $\lambda > 0, \mu$.

This lemma leads easily to the following result.

LEMMA 1.2.3: *Suppose that* $\{M_n\}$ *is any sequence of* r.v.'s *(not necessarily maxima) and* $\{a_n > 0\}$, $\{b_n\}$, *are real constants such that for each* $k = 1, 2, \ldots,$

$$P^k\{a_{kn}(M_n - b_{kn}) \leqslant x\} \to G(x) \qquad \text{as } n \to \infty \qquad (1.2.1)$$

(*where* G *is a nondegenerate* d.f.). *Then corresponding to each* $k = 1, 2, \ldots$, *there are constants* $\alpha_k > 0$, β_k *such that*

$$G^k(\alpha_k x + \beta_k) = G(x). \tag{1.2.2}$$

Proof: If G_n is the d.f. of $a_n(M_n - b_n)$, $G_n(x) \to G(x)$ and it is easily checked from (1.2.1) that for any given fixed k, $G_n(\lambda_n x + \mu_n) \to G^{1/k}(x)$, where $\lambda_n = a_n/a_{kn}$, $\mu_n = a_n(b_{nk} - b_n)$. Since $G^{1/k}(x)$ is a nondegenerate d.f. it must (by Lemma 1.2.2) be equal to $G(\alpha_k x + \beta_k)$ for some $\alpha_k > 0$, β_k, as required.

LEMMA 1.2.4: *Let* G *be a nondegenerate* d.f. *such that* (1.2.2) *holds for each* $k = 1, 2, \ldots$. *Then* G *is one of the three extreme value types being Type I, II, or III according as* $\alpha_k = 1$ *for some* k, $\alpha_k > 1$ *for some* k, $\alpha_k < 1$ *for some* k. (*Hence, e.g., if* $\alpha_k = 1$ *for some* k *then* $\alpha_k = 1$ *for all* k, *and similarly for* $\alpha_k < 1$, $\alpha_k > 1$.)

The derivation of this result (which we shall not give here) forms the major part of Gnedenko's proof in [16].

By looking at Lemmas 1.2.3 and 1.2.4, we see that if (1.2.1) holds for all $k = 1, 2, \ldots$, then G must be one of the three extreme value types. But in the statement of Gnedenko's Theorem, (1.2.1) is *assumed* for $k = 1$. Hence *the theorem will follow if it can be shown that the truth of* (1.2.1) *for* $k = 1$ *implies its truth for all* k. This may be shown trivially (Lemma 1.2.5) in the case where the ξ_n are i.i.d. However, it holds more generally, for dependent ξ_n (e.g., under the weak mixing condition of the next chapter). Since the i.i.d. assumption does not appear in the above lemmas at all, we see that *Gnedenko's theorem will follow for any class of sequences* $\{\xi_n\}$*—not necessarily* i.i.d.*—such that the truth of* (1.2.1) *for* $k = 1$ *implies its truth for* $k = 2, 3, 4, \ldots$. We shall return to this and pursue its consequences further in Chapter 2. For the i.i.d. case considered here the following simple result thus completes the proof of Gnedenko's Theorem.

LEMMA 1.2.5: *If* $\{\xi_n\}$ *is a sequence of* i.i.d. *random variables* (*with common* d.f. F) *and if* (1.2.1) *holds for* $k = 1$, *then* (1.2.1) *holds for all*

$k = 1, 2, 3 \ldots$. *Hence Gnedenko's Theorem holds for* i.i.d. *sequences by the above lemmas and remarks.*

Proof:
$$\begin{aligned} P^k\{a_{kn}(M_n - b_{kn}) \leqslant x\} &= P^k\{M_n \leqslant x/a_{kn} + b_{kn}\} \\ &= F^{nk}(x/a_{kn} + b_{kn}) \\ &= P\{M_{kn} \leqslant x/a_{kn} + b_{kn}\} \\ &= P\{a_N(M_N - b_N) \leqslant x\} \end{aligned}$$

with $N = kn$. But this tends to $G(x)$ as n (and hence N) $\to \infty$ so that (1.2.1) follows for all k, as required.

It is of course of interest to know which (if any) of the three possible limiting forms corresponds to a given sequence $\{\xi_n\}$ of i.i.d. random variables, i.e., to which *domain of attraction* the common d.f. of the ξ_n belongs. First of all we note that there are cases where *no* nondegenerate limit exists whatever a_n, b_n are chosen. This is easily proved for example if $F(x_0-) < F(x_0) = 1$ for some finite x_0 (i.e., if the distribution of the ξ_n has a finite right endpoint with a jump there). General conditions under which no nondegenerate limit exists are given in [16] and include for example the case where $F(x)$ is a *Poisson* d.f.

On the other hand, various necessary and sufficient conditions are known for F to belong to any given one of the three possible domains of attraction. While we shall not need these explicitly in later sections, we state, for reference, one such set of necessary and sufficient conditions for the d.f. F of the i.i.d. sequence $\{\xi_n\}$ to belong to each domain of attraction.

Type I: There exists a continuous function $A(x)$ such that $\lim_{x \uparrow x_0} A(x) = 0$ where x_0 ($\leqslant \infty$) is such that $F(x_0) = 1$ and $F(x) < 1$ for $x < x_0$, and such that for all h, $\lim_{x \uparrow x_0}\{[1 - F\{x(1 + hA(x))\}]/[1 - F(x)]\} = e^{-h}$.

Type II: $\lim_{x \to \infty}\{[1 - F(x)]/[1 - F(kx)]\} = k^\alpha, \alpha > 0$, each $k > 0$.

Type III: There exists x_0 (finite) such that $x_0 = \sup\{x: F(x) < 1\} < \infty$ and $\lim_{h \downarrow 0}\{[1 - F(x_0 - kh)]/[1 - F(x_0 - h)]\} = k^\alpha$, each $k > 0$.

It is easily checked from these criteria that each extreme value d.f. belongs to its own domain of attraction.

The case when the ξ_i are normal is of special interest. It is rather easy to show by taking $x_0 = \infty$, $A(x) = 1/x^2$ from the above criteria that the normal distribution belongs to the Type I domain of attraction. Of course this does not give the normalizing constants a_n, b_n. However it will be instructive to indicate how the (Type I) limiting distribution in the normal case (and the normalizing constants) may be obtained directly without recourse to the above criteria. To do this (and for other purposes) we need the following result.

LEMMA 1.2.6: *If* $\{\xi_n\}$ *is an* i.i.d. *sequence* (d.f. F), $\tau \geqslant 0$ *and if* $\{u_n\}$ *is a sequence of real numbers such that*

$$1 - F(u_n) = \tau/n + o(1/n) \qquad \text{as } n \to \infty, \tag{1.2.3}$$

then

$$P(M_n \leqslant u_n) \to e^{-\tau} \qquad \text{as } n \to \infty \tag{1.2.4}$$

where $M_n = \max(\xi_1, \xi_2, \ldots, \xi_n)$. *Conversely if* (1.2.4) *holds for some* $\tau \geqslant 0$, *so does* (1.2.3).

Proof: $P\{M_n \leqslant u_n\} = F^n(u_n) = [1 - (1-F(u_n))]^n$ so that (1.2.3) at once implies (1.2.4). Conversely if (1.2.4) holds it is clear that $F(u_n) \to 1$ and by the above that

$$n \log[1 - (1 - F(u_n))] \to -\tau$$

which easily yields (1.2.3).

We note in passing that, unless F is continuous, it is not necessarily possible to choose a sequence $\{u_n\}$ to satisfy (1.2.3) (as may be shown, e.g., for the Poisson distribution). In such a case clearly no (normalized) limiting distribution for M_n exists.

Let us turn now to the determination of the limiting distribution and the sequences of normalizing constants in cases where the limiting distribution does exist. Suppose we can find constants $a_n > 0$, b_n such that for each real x, writing $u_n = x/a_n + b_n$

$$n(1 - F(u_n)) \to \tau = \tau(x). \tag{1.2.5}$$

Then by Lemma 1.2.6, $P\{M_n \leqslant u_n\} \to e^{-\tau}$ or

$$P\{a_n(M_n - b_n) \leqslant x\} \to e^{-\tau(x)} = G(x), \text{ say.}$$

Hence if $\{a_n\}$, $\{b_n\}$ are found such that (1.2.5) holds with $u_n = x/a_n + b_n$ for each x, then these are the appropriate normalizing constants and, moreover, the limiting d.f. $G(x) = e^{-\tau(x)}$. As an important example we indicate the use of this procedure in the normal case. Specifically if $\xi_1, \xi_2, \ldots,$ are i.i.d. standard normal r.v.'s and $\tau > 0$, we seek u_n of the form $x/a_n + b_n$ such that (1.2.5) holds, i.e.,

$$1 - \Phi(u_n) \sim \tau/n$$

where Φ denotes the standard normal d.f. Since, as $u \to \infty$, $1 - \Phi(u) \sim \phi(u)/u$ (where $\phi(u) = (2\pi)^{-1/2} e^{-u^2/2}$) we require $n\phi(u_n)/u_n \to \tau$ or, taking logarithms, that

$$u_n^2/2 + \log u_n - \log n + \tfrac{1}{2}\log 2\pi + \log \tau \to 0. \qquad (1.2.6)$$

From (1.2.6) it follows that we must have $u_n^2/(2\log n) \to 1$ giving $\log u_n = \frac{1}{2}(\log 2 + \log\log n + o(1))$, which, inserted in (1.2.6) gives, by standard calculations,

$$u_n = (2\log n)^{1/2}\left[1 - \frac{\log\tau + \frac{1}{2}\log 4\pi + \frac{1}{2}\log\log n}{2\log n} + o\left(\frac{1}{\log n}\right)\right]. \qquad (1.2.7)$$

Thus if we choose

$$\begin{cases} a_n = (2\log n)^{1/2} \\ b_n = (2\log n)^{1/2} - \frac{1}{2}(2\log n)^{-1/2}(\log\log n + \log 4\pi) \end{cases} \qquad (1.2.8)$$

we must require, to satisfy (1.2.5)

$$u_n = \frac{-\log\tau}{a_n} + b_n + o(\log n)^{-1/2}.$$

It is now easily checked that if we simply take $u_n = -(\log \tau/a_n) + b_n$ then (1.2.6) holds and so does (1.2.5). Writing $\tau(x) = e^{-x}$ we obtain $u_n = x/a_n + b_n$ as required, and the limiting d.f. $G(x) = e^{-\tau(x)} = \exp(-e^{-x})$. For reference we summarize this as a theorem.

THEOREM 1.2.7: *If* $\xi_1, \xi_2, \ldots,$ *are* i.i.d. *standard normal* r.v.'s *and* $M_n = \max\{\xi_1, \xi_2, \ldots, \xi_n\}$, *then*

$$P\{a_n(M_n - b_n) \leqslant x\} \to \exp(-e^{-x}) \qquad \text{as } n \to \infty,$$

where a_n, b_n *are given by* (1.2.8).

1.3. Distribution of kth largest values—fixed rank k. For the remainder of this chapter on the classical theory we look at the kth largest $M_n^{(k)}$ of $\xi_1, \ldots, \xi_n$. In this section we shall consider (in reasonable detail) the case where k is fixed, and in the remaining sections indicate the corresponding results when $k \to \infty$ with n.

It will be convenient to first state the following lemma from which Lemma 1.2.6 may be easily generalized.

LEMMA 1.3.1: *Let k be a fixed non-negative integer,* $\tau > 0$, *and* $\{p_n\}$ *a sequence with* $0 < p_n < 1$. *Then*

$$\sum_{j=0}^{k} \binom{n}{j} p_n^j (1 - p_n)^{n-j} \to e^{-\tau} \sum_{j=0}^{k} \tau^j/j! \tag{1.3.1}$$

if and only if $np_n \to \tau$. *Hence also if* (1.3.1) *holds for any one non-negative integer k it holds for all such k.*

Indication of Proof: If $np_n \to \tau$, (1.3.1) is simply the well-known Poisson limit for the binomial distribution (and is easily proved directly). The converse is also easy to show, though some details are required (and these are given in full in [23]). The steps are

(i) to show from (1.3.1) that

$$\sum_{j=0}^{k} \frac{n^j}{j!} p_n^j (1 - p_n)^{n-j} \to e^{-\tau} \sum_{j=0}^{k} \tau^j/j!$$

(ii) to note that this requires that np_n be bounded and further, then, that

$$e^{-np_n} \sum_{j=0}^{k} \frac{(np_n)^j}{j!} \to e^{-\tau} \sum_{j=0}^{k} \tau^j/j!$$

(iii) to use the continuity of the inverse of the function $\psi(x) = e^{-x} \sum_{j=0}^{k} x^j/j!$ to show that $np_n \to \tau$.

Note that if (1.3.1) is assumed to hold for all $k = 1, 2, 3, \ldots$, then it follows from standard convergence criteria that $np_n \to \tau$, or indeed this follows directly from the case $k = 0$. The novelty here is that (1.3.1) is assumed to hold for *any* one fixed k (though as noted in the lemma, it must then hold for all such k).

Armed with this, we may obtain the following immediate generalization of Lemma 1.2.6. In this, S_n will denote the number of exceedances of u_n by $\xi_1, \ldots, \xi_n$, i.e., the number of i, $1 \leqslant i \leqslant n$, such that $\xi_i > u_n$.

LEMMA 1.3.2: *Let $\xi_1, \xi_2, \ldots,$ be an* i.i.d. *sequence of* r.v.'s *with common* d.f. F, $\tau \geqslant 0$, *and let $\{u_n\}$ be a real sequence such that* (1.2.3) *holds, i.e.,*

$$1 - F(u_n) = \tau/n + o(1/n)$$

then

$$P\{S_n \leqslant k\} \to e^{-\tau} \textstyle\sum_{j=0}^{k} \tau^j/j! \tag{1.3.2}$$

for all $k = 0, 1, 2, 3, \ldots$. Conversely if (1.3.2) *holds for any one fixed nonnegative integer k then* (1.2.3) *holds, and* (1.3.2) *holds for all such k.*

Proof: $S_n = \sum_1^n \chi_i$ where $\chi_i = 1$ if $\xi_i > u_n$ and $\chi_i = 0$ if $\xi_i \leqslant u_n$. Since $P(\chi_i = 1) = 1 - F(u_n)$, S_n is binomial with parameters n, $p_n = 1 - F(u_n)$, and thus

$$P\{S_n \leqslant k\} = \sum_{j=0}^{k} \binom{n}{j} p_n^j (1 - p_n)^{n-j}.$$

Lemma 1.3.1 now gives all the conclusions stated.

The case $k = 0$ is of course that covered by Lemma 1.2.6. (The events $S_n = 0$, $M_n \leqslant u_n$ are the same.)

Lemma 1.3.2 shows that the number S_n of exceedances of the level u_n by $\xi_1, \xi_2, \ldots, \xi_n$ has an asymptotic Poisson distribution. It should be noted that the mean number of exceedances is $n(1 - F(u_n)) \to \tau$, i.e., the "level" u_n is "coordinated" with the number n to make this expectation finite. It is also of interest to note that the exceedances of the level u_n themselves take on the character of a Poisson process when n becomes large (subject to an appropriate change of "time scale"). This topic will be discussed later under more general conditions.

Our immediate purpose here is to show how Gnedenko's Theorem may be generalized to deal with the kth largest $M_n^{(k)}$ of the i.i.d. random variables $\xi_1, \xi_2, \ldots, \xi_n$.

THEOREM 1.3.3: *Let* $\xi_1, \xi_2, \ldots,$ *be* i.i.d. (*with* d.f. F) *and suppose that*

$$P\{a_n(M_n - b_n) \leqslant x\} \to G(x) \tag{1.3.3}$$

for some nondegenerate (and hence Type I, II, or III) d.f. G. *Then for any fixed positive integer* k

$$P\{a_n(M_n^{(k)} - b_n) \leqslant x\} \to G(x) \sum_{j=0}^{k-1} (-\log G(x))^j/j! \tag{1.3.4}$$

(*interpreted as zero where* $G(x) = 0$), *with the same* a_n, b_n. *Conversely if* $P\{a_n(M_n^{(k)} - b_n) \leqslant x\} \to H(x)$ *for some nondegenerate* d.f. $H(x)$ (*and any one fixed* k), *then* $H(x)$ *has the form on the right of* (1.3.4) *where* G *satisfies* (1.3.3), (*and hence* (1.3.4) *holds for all* k).

Proof: If (1.3.3) holds, then (where $G(x) > 0$) (1.2.4) holds with $u_n = x/a_n + b_n$, $\tau = -\log G(x)$. Hence by Lemma 1.2.6, (1.2.3) holds and Lemma 1.3.2 gives (1.3.2) for all k. Putting $k - 1$ for k in (1.3.2) we obtain (1.3.4) since the events $\{M_n^{(k)} \leqslant u_n\}$, $\{S_n \leqslant k - 1\}$ are the same.

In the other direction, if (1.3.4) holds then so does (1.3.2) with $(k - 1)$ replacing k. Hence by Lemma 1.3.2, (1.3.2) holds for all k and in particular $k = 0$, which is just (1.3.3).

Thus we have the satisfactory situation that the nondegenerate limiting laws for $M_n^{(k)}$ are all of the form (1.3.4), based on the same extreme value d.f. $G(x)$ as M_n, and using the same normalizing constants a_n, b_n.

1.4. Variable ranks. For fixed k, the available information regarding the asymptotic distribution of the kth largest, $M_n^{(k)}$, is satisfying and rather complete. The information is also rather complete [36] when we allow k to tend to infinity at the same rate as n, i.e., consider $M_n^{(k_n)}$ where $k_n \to \infty$, $k_n/n \to \theta$ (cf. section 1.5). When $k_n \to \infty$, in more general ways the picture is not quite so tidy. However, many results are known, and we indicate some of these in this section. First we note the following counterparts of Lemmas 1.3.1 and 1.3.2.

LEMMA 1.4.1: *Let $\{k_n\}$ be any sequence of positive integers and $\{p_n\}$, $(0 < p_n < 1)$ such that $np_n(1 - p_n) \to \infty$. Then*

$$\sum_{j=0}^{k_n} \binom{n}{j} p_n^j (1 - p_n)^{n-j} \to \Phi(\lambda) \tag{1.4.1}$$

(*where Φ is the standard normal* d.f. *and λ any fixed constant*) *if and only if*

$$\frac{k_n - np_n}{\sqrt{np_n(1 - p_n)}} \to \lambda. \tag{1.4.2}$$

Equivalently this holds if and only if

$$k_n \to \infty, \qquad n - k_n \to \infty, \qquad \text{and} \qquad \frac{k_n - np_n}{\sqrt{k_n(1 - k_n/n)}} \to \lambda. \tag{1.4.3}$$

Indication of Proof: If $\chi_1, \ldots, \chi_n$ are i.i.d., $P(\chi_i = 1) = p_n = 1 - P(\chi_i = 0)$, it follows from the Berry-Esseen bound that

$$P\left\{\frac{\sum_1^n \chi_i - np_n}{\sqrt{np_n(1 - p_n)}} \leqslant \frac{k_n - np_n}{\sqrt{np_n(1 - p_n)}}\right\} - \Phi\left\{\frac{k_n - np_n}{\sqrt{np_n(1 - p_n)}}\right\}$$

$$\leqslant \frac{C}{\sqrt{np_n(1 - p_n)}}$$

which tends to zero since $np_n(1 - p_n) \to \infty$. The first term on the left is just

$$P\left\{\sum_1^n \chi_i \leqslant k_n\right\} = \sum_{j=0}^{k_n} \binom{n}{j} p_n^j (1 - p_n)^{n-j}$$

and the first result follows since

$$\Phi\left(\frac{k_n - np_n}{\sqrt{np_n(1 - p_n)}}\right) \to \Phi(\lambda)$$

if and only if

$$\left(\frac{k_n - np_n}{\sqrt{np_n(1 - p_n)}}\right) \to \lambda.$$

That (1.4.2) implies (1.4.3) follows by writing

$$k_n = np_n + \lambda\sqrt{np_n(1 - p_n)}(1 + o(1))$$

and noting that this implies $k_n \sim np_n$ and $(n - k_n) \sim n(1 - p_n)$. Similarly (1.4.3) implies (1.4.2).

LEMMA 1.4.2: *If* $k_n \to \infty$, $n - k_n \to \infty$, *and if* u_n *satisfies*

$$\frac{k_n - n(1 - F(u_n))}{\sqrt{k_n\,(1 - k_n/n)}} \to \lambda \tag{1.4.4}$$

then

$$P\{S_n \leqslant k_n\} \to \Phi(\lambda) \tag{1.4.5}$$

(*where* S_n *is again the number of* $\xi_1, \ldots, \xi_n$ *which exceed* u_n). *Conversely, if* (1.4.5) *holds, so does* (1.4.4).

This follows from Lemma 1.4.1 by writing $p_n = 1 - F(u_n)$.

Equation (1.4.5) may be restated to give the limit for $P\{M_n^{(k_n)} \leqslant u_n\}$ $(=P\{S_n \leqslant k_n - 1\}) \to \Phi(\lambda)$ under the conditions of the lemma. In particular, the following result of Smirnov [36] holds.

THEOREM 1.4.3: *Suppose there exist constants $a_n > 0$, b_n such that*

$$\frac{k_n - n[1 - F(x/a_n + b_n)]}{\sqrt{k_n(1 - k_n/n)}} \to \lambda(x) \qquad -\infty < x < \infty \tag{1.4.6}$$

where $k_n \to \infty$, $n - k_n \to \infty$. Then

$$P\{a_n(M_n^{(k_n)} - b_n) \leqslant x\} \to \Phi(\lambda(x)). \tag{1.4.7}$$

Conversely, if (1.4.7) *holds so does* (1.4.6).

Sometimes it is convenient to use (1.4.6) to find the constants a_n, b_n and the limit $\lambda(x)$ leading to the limiting law (1.4.6), in an analogous way to that used previously for the maximum.

Put in another way we see that

$$P\{a_n(M_n^{(k_n)} - b_n) \leqslant x\} \to G(x) \tag{1.4.8}$$

if and only if (1.4.6) holds with $G(x) = \Phi(\lambda(x))$. A question of great interest, of course, is what d.f.'s G of this form are possible (cf. Gnedenko's Theorem for the maximum, and Theorem 1.3.3). Such a study is most naturally divided into two parts:

(a) when $k_n \to \infty$, $k_n/n \to \theta \qquad 0 < \theta < 1$,

(b) $k_n \to \infty$, $k_n/n \to 0$, i.e., $k_n = o(n)$.

(Corresponding results are of course obtainable when k_n is replaced by $n - k_n$—i.e., considering smallest instead of largest values.) We refer to Case (a) as that of *central* ranks, and (b) as the case of *intermediate* ranks. We deal briefly with these cases separately in the following two sections.

1.5. Central ranks. The case of central ranks, where $k_n/n \to \theta$ $(0 < \theta < 1)$, has been studied extensively in [36]. While we shall have little to say about this case in later sections, a few basic facts for the i.i.d. sequence will be discussed here for completeness. First we note that it is possible for two sequences $\{k_n\}$, $\{k'_n\}$ with $\lim k_n/n = \lim k'_n/n$ to lead to *different* nondegenerate limiting d.f.'s for $M_n^{(k_n)}$,

$M_n^{(k_n')}$. Specifically (as shown in [36]), we may have $k_n/n \to \theta$, $k_n'/n \to \theta$ and

$$P\{a_n(M_n^{(k_n)} - b_n) \leqslant x\} \to G(x), \tag{1.5.1}$$

$$P\{a_n'(M_n^{(k_n')} - b_n') \leqslant x\} \to G'(x), \tag{1.5.2}$$

where $a_n > 0$, b_n, $a_n' > 0$, b_n' are constants and $G(x)$, $G'(x)$ are nondegenerate d.f.'s of different "type." On the other hand, this is precluded under the requirement that

$$\sqrt{n}\left(\frac{k_n}{n} - \theta\right) \to 0 \tag{1.5.3}$$

as the following result shows.

LEMMA 1.5.1: *Suppose that* (1.5.1) *and* (1.5.2) *hold where G and G' are nondegenerate, and $\{k_n\}$, $\{k_n'\}$ both satisfy* (1.5.3). *Then G and G' are of the same "type," i.e., $G'(x) = G(ax + b)$ for some a, b.*

Proof: If F is the d.f. of the terms of the i.i.d. sequence $\xi_1, \xi_2, \ldots$, and (1.5.1) holds, then, by Theorem 1.4.3,

$$\frac{k_n - n[1 - F(x/a_n + b_n)]}{\sqrt{k_n(1 - k_n/n)}} \to \lambda(x)$$

where $G(x) = \Phi(\lambda(x))$. By (1.5.3) this implies

$$\sqrt{n}\,\frac{\theta - [1 - F(x/a_n + b_n)]}{\sqrt{\theta(1 - \theta)}} \to \lambda(x).$$

By (1.5.3) with k_n' for k_n we can see therefore that

$$\frac{k_n' - n[1 - F(x/a_n + b_n)]}{\sqrt{k_n'(1 - k_n'/n)}} \to \lambda(x)$$

so that, by Theorem 1.4.3 again,

$$P\{a_n(M_n^{(k)'_n} - b_n) \leqslant x\} \to \Phi(\lambda(x)) = G(x).$$

If G_n denotes the d.f. of $M_n^{(k_n')}$, this says that $G_n(x/a_n + b_n) \to G(x)$, whereas, by (1.5.2), $G_n(x/a_n' + b_n') \to G'(x)$. Hence by Lemma 1.2.2, G and G' are of the same type.

If the sequence $\{k_n\}$ satisfies (1.5.3), then there are just four types of limiting distributions for $M_n^{(k_n)}$, i.e., for possible forms of nondegenerate G in (1.5.1). We state this as a theorem for completeness and refer the reader to [36] for proof, as well as a complete characterization of domains of attraction.

THEOREM 1.5.2: *If the central rank sequence* $\{k_n\}$ *satisfies* (1.5.3) *the only possible nondegenerate* d.f.'s G *for which* (1.5.1) *holds are*

$$
\begin{array}{llll}
1. & G(x) = 0, & x < 0 & \\
 & \quad\;\; = \Phi(cx^{\alpha}), & x \geqslant 0 & (c > 0,\ \alpha > 0) \\
2. & G(x) = \Phi(-c|x|^{\alpha}), & x < 0 & (c > 0,\ \alpha > 0) \\
 & \quad\;\; = 1, & x \geqslant 0 & \\
3. & G(x) = \Phi(-c_1|x|^{\alpha}), & x < 0 & \\
 & \quad\;\; = \Phi(c_2 x^{\alpha}), & x \geqslant 0 & (c_1 > 0,\ c_2 > 0,\ \alpha > 0) \\
4. & G(x) = 0, & x < -1 & \\
 & \quad\;\; = \tfrac{1}{2}, & -1 \leqslant x < 1 & \\
 & \quad\;\; = 1, & x > 1. &
\end{array}
$$

1.6. Intermediate ranks. By an *intermediate* rank sequence we mean a sequence $\{k_n\}$ such that $k_n \to \infty$ but $k_n = o(n)$. The theory of Section 1.4 applies, with some slight simplification. For example the criterion (1.4.3) may be rephrased as

$$P_n = \frac{k_n}{n} - \lambda\frac{\sqrt{k_n}}{n} + o\left(\frac{\sqrt{k_n}}{n}\right)$$

and correspondingly (1.4.4) has this same form with $1 - F(u_n)$ for P_n. The following result (of [43]) gives the possible limiting d.f.'s for $M_n^{(k_n)}$ when k_n is nondecreasing.

THEOREM 1.6.1: *If* $\xi_1, \xi_2, \ldots,$ *are* i.i.d. *and* $\{k_n\}$ *is a nondecreasing intermediate rank sequence, and if there are constants* $a_n > 0$, b_n *such that*

$$P\{a_n(M_n^{(k_n)} - b_n) \leqslant x\} \to G(x)$$

for a nondegenerate d.f. G, *then* G *has one of the three forms*

$$\begin{aligned} G_1(x) &= \Phi(-a \log |x|), & x &< 0 & (a > 0) \\ &= 1, & x &\geqslant 0 & \\ G_2(x) &= 0, & x &\leqslant 0 & (a > 0) \\ &= \Phi(a \log x), & x &> 0 & \\ G_3(x) &= \Phi(x), & -\infty &< x < \infty. & \end{aligned}$$

Theorem 1.6.1 is rather satisfying, though it does not specify the domains of attraction of the three limiting forms. Some results in this direction have been obtained in [37], [43], [9]. However these are highly dependent on the rank sequence $\{k_n\}$. For example a class of rank sequences $\{k_n\}$ such that $k_n \sim l^2 n^\theta$ $(l > 0, 0 < \theta < 1)$ are studied in [9]. If F is any d.f. it is known that there is at most one (l, θ) pair such that F belongs to the domain of attraction of G_1 (and the same statement holds also for G_2). In addition, there are rank sequences such that only the normal law G_3 is a possible limit and, moreover, there are distributions attracted to it for every intermediate rank sequence $\{k_n\}$. For further details on this topic we refer to [37], [43], [9].

This completes our brief survey of selected topics of classical extreme value theory. While obviously far from being comprehensive, these topics have been chosen to illustrate the theory, and our task in the next chapters will be to indicate the extent to which at least quite a number of properties of this type apply in appropriate dependent situations.

2. EXTREME VALUE THEORY FOR STATIONARY SEQUENCES

2.1. Mixing and related conditions. As noted in Chapter 1, to include dependence we shall now turn to a consideration of *stationary* sequences $\{\xi_n : n = 1, 2, \ldots\}$ as being a natural generalization of i.i.d. cases. Precisely, by stationarity we mean that the joint distributions of $(\xi_{i_1}, \xi_{i_2}, \ldots, \xi_{i_n})$ and of $(\xi_{i_1+m}, \xi_{i_2+m}, \ldots, \xi_{i_n+m})$ are the same for any choice of positive integers $n, i_1, i_2, \ldots, i_n, m$.

Stationarity requires that the "statistical properties" of the sequence $\{\xi_n\}$ do not change with "time, n." We shall further require certain restrictions on the *dependence* of the ξ_n-sequence. Next to independence itself, the strongest such restriction would be that of so-called "m-dependence," which allows ξ_i and ξ_j to be dependent when $|i - j| \leqslant m$, but requires them (in particular) to be independent for $|i - j| > m$. More precisely, if for any $i_1, i_2, \ldots, i_n$ the joint d.f. of $\xi_{i_1}, \ldots, \xi_{i_n}$ is

$$F_{i_1, \ldots, i_n}(x_1, \ldots, x_n) = P\{\xi_{i_1} \leqslant x_1 \cdots \xi_{i_n} \leqslant x_n\}$$

then m-dependence requires that

$$F_{i_1, \ldots, i_p, j_1 \ldots, j_q}(x_1, \ldots, x_p, y_1, \ldots, y_q)$$
$$= F_{i_1, \ldots, i_p}(x_1, \ldots, x_p) F_{j_1, \ldots, j_q}(y_1, \ldots, y_q) \qquad (2.1.1)$$

whenever $i_1, \ldots, i_p, j_1, \ldots, j_q$ is any choice of integers such that $i_1 < i_2 \cdots < i_p < j_1 \cdots < j_q$, $j_1 - i_p > m$ (and the $x_1, \ldots, x_p$, $y_1, \ldots, y_q$ are any real values).

An obvious weakening of m-dependence is to require that the difference between the two sides of (2.1.1) be appropriately small—but not necessarily zero—when the "separation" m is large. Further, the fact that, if $M_n = \max(\xi_1, \xi_2, \ldots, \xi_n)$,

$$P\{M_n \leqslant u\} = F_{1,2, \ldots, n}(u, u, \ldots, u)$$

suggests that we might well consider a condition in which the values $x_1, \ldots, x_p$, $y_1, \ldots, y_q$ are all equal—ignoring all other possible values. With this in mind we write, for brevity,

$$F_{i_1, \ldots, i_n}(u, u, \ldots, u) = F_{i_1, \ldots, i_n}(u). \qquad (2.1.2)$$

Then we shall say that *the Condition D holds* if for any choice of $i_1 < i_2 \cdots < i_p < j_1 \ldots < j_q, j_1 - i_p \geqslant l$, we have

$$|F_{i_1, \ldots, i_p, j_1, \ldots, j_q}(u) - F_{i_1, \ldots, i_p}(u) F_{j_1, \ldots, j_q}(u)| \leqslant \alpha_l \qquad (2.1.3)$$

where $\alpha_l \to 0$ as $l \to \infty$.

While Condition D is a natural weakening of m-dependence for consideration of extreme values, we shall be able to use a still weaker condition, as will be indicated below. However, we first note, in passing, the relationship of Condition D to the "strong mixing" condition often used. By *strong mixing* we mean (cf. [34]) that there is a sequence $\alpha_l \to 0$ as $l \to \infty$ such that

$$|P(A \cap B) - P(A)P(B)| < \alpha_l$$

whenever $A \in \sigma\{\xi_1, \ldots, \xi_p\}$, $B \in \sigma(\xi_{p+l+1}, \xi_{p+l+2}, \ldots)$ for any p and l, $\sigma\{\cdot\}$ denoting the σ-fields generated by the indicated r.v.'s. It is easily seen that m-dependence implies strong mixing. But strong mixing clearly implies our condition D—which may be seen at once by taking A, B to be the events $\{\xi_{i_1} \leqslant u \cdots \xi_{i_p} \leqslant u\}$ and $\{\xi_{j_1} \leqslant u \cdots \xi_{j_q} \leqslant u\}$, respectively.

As noted above, while Condition D is itself rather weak, we shall in fact find a significantly weaker condition yet, to be the most useful. Specifically, if $\{u_n\}$ is a sequence of real numbers, we shall say that the *Condition* $D(u_n)$ *is satisfied* if for any choice of n, $i_1 < i_2 \cdots < i_p < j_1 < j_2 \cdots < j_q$, $j_1 - i_p \geqslant l$ we have

$$|F_{i_1,\ldots i_p, j_1, \ldots, j_q}(u_n) - F_{i_1, \ldots, i_p}(u_n)F_{j_1, \ldots, j_q}(u_n)| \leqslant \alpha_{n,l} \tag{2.1.4}$$

where $\alpha_{n,l}$ is nonincreasing in l and where $\lim_{n\to\infty} \alpha_{n,l_n} = 0$ for some sequence $l_n \to \infty$ and such that $l_n/n \to 0$. (Note that if (2.1.4) holds for some $\alpha_{n,l}$ it is clearly possible to take $\alpha_{n,l}$ to be nonincreasing in l.)

Obviously Condition D implies $D(u_n)$ for *any* sequence $\{u_n\}$. In practice, however, we shall only require $D(u_n)$ to hold for certain specific types of sequences (in fact sequences satisfying (1.2.3)). We now give a lemma which provides a useful form for the application of $D(u_n)$. If E is any set of integers, $M(E)$ will denote $\max\{\xi_j : j \in E\}$, ($M(E) = M_n$ if $E = (1, 2, \ldots, n)$). By an *interval* we shall mean any finite set E of consecutive integers $(j, j+1, \ldots, l)$ say. Two sets E_1, E_2 will be called *separate* if they are subsets of disjoint intervals. If the largest member of E_1 is l_1 and the smallest member of E_2 is j_2, with $j_2 > l_1$, the *separation* of E_1 and E_2 is $j_2 - l_1$.

LEMMA 2.1.1: *Suppose $D(u_n)$ holds for some sequence $\{u_n\}$. Let N, r, l be fixed integers and let $E_1, E_2, \ldots, E_r$ be sets such that E_i and E_j are separated by at least l when $i \neq j$. Then*

$$\left|P\left(\bigcap_{i=1}^{r} \{M(E_i) \leqslant u_N\}\right) - \prod_{i=1}^{r} P\{M(E_i) \leqslant u_N\}\right| \leqslant (r-1)\alpha_{N,l}. \tag{2.1.5}$$

Proof: Let $E_1, E_2, \ldots, E_r$ be renumbered if necessary in order of increasing size of elements (i.e., E_1 contains the smallest elements, etc.). The statement of the result for $r = 2$ is just $D(u_n)$. The entire result is then a simple induction from the fact that the left-hand side of (2.1.5) is dominated by

$$\left|P\left(\{M(E_1) \leqslant u_N\} \cap \bigcap_2^r \{M(E_i) \leqslant u_N\}\right)\right.$$

$$\left. - P\{M(E_1) \leqslant u_N\} P\left(\bigcap_2^r \{M(E_i) \leqslant u_N\}\right)\right|$$

$$+ P\{M(E_1) \leqslant u_N\} \left|P\left(\bigcap_2^r \{M(E_i) \leqslant u_N\}\right) - \prod_2^r P\{M(E_i) \leqslant u_N\}\right|$$

$$\leqslant \alpha_{N,l} + (r - 2)\alpha_{N,l} = (r - 1)\alpha_{N,l}$$

using $D(u_n)$ for the first term, and assuming the result true for $(r - 1)$ sets, for the second.

We shall see that Gnedenko's Theorem can be generalized to apply to stationary sequences satisfying conditions of the form $D(u_n)$. For other results we shall need another condition which restricts the bivariate distribution of ξ_i and ξ_j at "high levels." Specifically, if $\{u_n\}$ is any real sequence we shall say that $D'(u_n)$ is satisfied by the sequence $\{\xi_n\}$ if

$$\limsup_{n \to \infty} \left[n \sum_{j=2}^{n} P\{\xi_1 > u_{nk}, \xi_j > u_{nk}\}\right] = o(1/k) \quad \text{as } k \to \infty. \quad (2.1.6)$$

It is easily shown that if $\{\xi_n\}$ are i.i.d. and if $\{u_n\}$ satisfies (1.2.3), then $D'(u_n)$ holds (the left-hand side of (2.1.6) being asymptotically τ^2/k^2). However, $D'(u_n)$ will rule out cases where "nearby" ξ_j's are too highly dependent. An example of such a case is that for which $\xi_n = \max(\eta_n, \eta_{n+1})$, where $\{\eta_n\}$ is an i.i.d. sequence. Then if $p_n = P(\eta_1 > u_n)$ it is easily checked that

$$P[\xi_1 > u_n] = p_n(2 - p_n)$$

$$P[\xi_1 > u_n, \xi_2 > u_n] = p_n[1 + p_n(1 - p_n)].$$

Hence if $\{u_n\}$ satisfies (1.2.3) we have $2p_n \sim \tau/n$ so that

$$n \sum_{j=2}^{n} P\{\xi_1 > u_{nk}, \xi_j > u_{nk}\} \geqslant nP\{\xi_1 > u_{nk}, \xi_2 > u_{nk}\} \to \tau/(2k),$$

so that $D'(u_n)$ is not satisfied when $\tau > 0$.

In this case the dependence between the events $\{\xi_1 > u_n\}, \{\xi_2 > u_n\}$ is too great (e.g., $\liminf P\{\xi_2 > u_n \mid \xi_1 > u_n\} > 0$) to permit even $\limsup_n nP\{\xi_1 > u_{nk}, \xi_2 > u_{nk}\}$ to tend to zero faster than k^{-1}. This example is useful, incidentally, in a variety of contexts.

2.2. Gnedenko's Theorem for stationary sequences. In this section Gnedenko's Theorem will be obtained for a class of stationary sequences satisfying distributional mixing conditions of the type $D(u_n)$. The broad method used is similar to that of the pioneering work [25] (done under strong mixing) and will be briefly described, followed by three lemmas in which the detailed facts required are listed, together with indications of proof. Full details of proof, if desired, may be found in [23]. First we state the theorem.

THEOREM 2.2.1: *Let $\{\xi_n\}$ be a stationary sequence and let $a_n > 0$, b_n be constants such that $P\{a_n(M_n - b_n) \leqslant x\}$ converges to a non-degenerate* d.f. $G(x)$. *Suppose that $D(u_n)$ is satisfied for $u_n = x/a_n + b_n$, for each real x. Then $G(x)$ has one of the three extreme value forms listed in Theorem* 1.2.1.

Note that it follows as a corollary that the theorem holds if the condition that $D(u_n)$ be satisfied by $u_n = x/a_n + b_n$ is replaced by the simpler (but more restrictive) condition D.

As noted in Chapter 1, to prove this result we need only show that the truth of (1.2.1) for $k = 1$ implies its truth for all k, i.e., that

$$P\{a_n(M_n - b_n) \leqslant x\} \to G(x) \tag{2.2.1}$$

implies

$$P^k\{a_{nk}(M_n - b_{nk}) \leqslant x\} \to G(x) \tag{2.2.2}$$

for each $k = 2, 3, \ldots$. This will clearly follow if we show that, writing $N = nk$ for each fixed $k = 2, 3, \ldots,$

$$P\{M_N \leqslant x/a_N + b_N\} - P^k\{M_n \leqslant x/a_N + b_N\} \to 0 \qquad \text{as } n \to \infty$$

(for if (2.2.1) holds, it holds also with $N = nk$ replacing n). Thus the theorem will certainly follow *if we show that*

$$P\{M_N \leqslant u_N\} - P^k\{M_n \leqslant u_N\} \to 0 \tag{2.2.3}$$

for any sequence $\{u_n\}$ such that $D(u_n)$ holds (for (2.2.3) will then hold in particular for the sequences $u_n = x/a_n + b_n$).

The method of proof of (2.2.3) is a standard argument (cf. [25]) in the use of mixing conditions. It involves the dividing of the interval $(1, 2, \ldots, N)$ into k intervals of length n, shortening these to separate them and using Lemma 2.1.1 to show approximate independence of the maxima on these intervals. Specifically we divide the first $N = nk$ integers into $2k$ intervals as follows. Let l be an integer (which will later tend to infinity, but more slowly than n) and write $I_1 = (1, 2, \ldots, n - l)$, $I_1^* = (n - l + 1, \ldots, n)$, $I_2 = (n + 1, \ldots, 2n - l)$, $I_2^* = (2n - l + 1, \ldots, 2n)$, and so on. $I_1, I_1^*, I_2, I_2^*, \ldots, I_k, I_k^*$ alternately contain $(n - l)$ and l members, and as n later increases the intervals $I_1, \ldots, I_k$ become much larger then $I_1^*, \ldots, I_k^*$.

The following lemma contains the main steps of the approximation. These are, broadly, to show that the "small" intervals I_j^* can essentially be disregarded, so that Lemma 2.1.1 can be applied to the (now separate) intervals $I_1, I_2, \ldots, I_k$.

LEMMA 2.2.2: *Let $\{u_n\}$ be a given sequence such that $D(u_n)$ holds for the stationary sequence $\{\xi_n\}$. Then, with the above notation,*

(i) $$0 \leqslant P\left(\bigcap_{j=1}^{k}\{M(I_j) \leqslant u_N\}\right) - P\{M_N \leqslant u_N\} \leqslant kP\{M(I_1) \leqslant u_N < M(I_1^*)\}$$

(ii) $$\left|P\left(\bigcap_{j=1}^{k}\{M(I_j) \leqslant u_N\}\right) - P^k\{M(I_1) \leqslant u_N\}\right| \leqslant k\alpha_{N,l}$$

(iii) $$|P^k\{M(I_1) \leqslant u_N\} - P^k\{M_n \leqslant u_N\} \leqslant kP\{M(I_1) \leqslant u_N < M(I_1^*)\}.$$

Hence, by combining (i), (ii), (iii),

$$|P\{M_N \leqslant u_N\} - P^k\{M_n \leqslant u_N\}|$$
$$\leqslant 2kP\{M(I_1) \leqslant u_N < M(I_1^*)\} + k\alpha_{N,l}. \quad (2.2.4)$$

Sketch of proof: (i) follows at once by noting that

$$\bigcap_{j=1}^{k} \{M(I_j) \leqslant u_N\} \supset \{M_N \leqslant u_N\}$$

and the difference between these implies that $M(I_j) \leqslant u_N < M(I_j^*)$ for some j (the latter events having probability independent of j by stationarity).

(ii) follows from Lemma 2.1.1 with I_j for E_j, noting that $P\{M(I_j) \leqslant u_N\}$ is independent of j.

(iii) may be obtained by noting that

$$0 \leqslant P\{M(I_1) \leqslant u_N\} - P\{M_n \leqslant u_N\} \leqslant P\{M(I_1) \leqslant u_N < M(I_1^*)\}$$

and (writing $y = P\{M(I_1) \leqslant u_N\}$, $x = P\{M_n \leqslant u_N\}$), using the fact that $0 \leqslant y^k - x^k \leqslant k(y - x)$ for $0 \leqslant x \leqslant y \leqslant 1$.

The next lemma gives a convenient bound for the right-hand side of (2.2.4).

LEMMA 2.2.3: *Under the conditions of Lemma 2.2.2, if $r \geqslant 1$ is any fixed integer, and n is sufficiently large, then*

$$P\{M(I_1) \leqslant u_N < M(I_1^*)\} \leqslant \frac{1}{r} + 2r\alpha_{N,l_N} \quad (2.2.5)$$

where $\{l_n\}$ is the sequence occurring in the statement of Condition $D(u_n)$.

Sketch of proof: Since $l_n = o(n)$ and k is fixed we may, for n sufficiently large, choose intervals $E_1, E_2, \ldots, E_r$, each containing l_N

members from $1, 2, \ldots, n - l_N$, so that they are separated from each other and from I_1^* by at least l_N. Then

$$P\{M(I_1) \leqslant u_N < M(I_1^*)\} \leqslant P\left(\{M(I_1^*) > u_N\} \cap \bigcap_{s=1}^{r} \{M(E_s) \leqslant u_N\}\right)$$

$$= P\left(\bigcap_{s=1}^{r} \{M(E_s) \leqslant u_N\}\right)$$

$$- P\left(\{M(I_1^*) \leqslant u_N\} \cap \bigcap_{s=1}^{r} \{M(E_s) \leqslant u_N\}\right).$$

Writing $P\{M(E_s) \leqslant u_N\} = p$ (independent of s by stationarity) and using Lemma 2.1.1, we may see that the two terms on the right differ from p^r and p^{r+1} (in absolute magnitude) by no more than $(r-1)\alpha_{N,l_N}$, $r\alpha_{N,l_N}$, respectively, so that the left-hand side of (2.2.5) does not exceed $p^r - p^{r+1} + 2r\alpha_{N,l_N}$, from which (2.2.5) follows, since $p^r - p^{r+1} \leqslant 1/(r+1)$.

LEMMA 2.2.4: *Let $\{u_n\}$ be a given sequence such that $D(u_n)$ holds for the stationary sequence $\{\xi_n\}$. Then* (2.2.3) *holds, i.e.,*

$$P\{M_N \leqslant u_N\} - P^k\{M_n \leqslant u_N\} \to 0 \qquad \textit{as } n \to \infty$$

for any fixed k ($N = nk$). Hence also Theorem 2.2.1 *is proved.*

Proof: By Lemma 2.2.3, for any fixed integer r

$$\limsup_n P\{M(I_1) \leqslant u_N < M(I_1^*)\} \leqslant \frac{1}{r}$$

and hence $\limsup_n P\{M(I_1) \leqslant u_N < M(I_1^*)\} = 0$. The result now follows from (2.2.4) on putting $l = l_N$, since both terms on the right then tend to zero.

2.3. Convergence of $P\{M_n \leqslant u_n\}$, and its consequences. One of the important results in the classical case is that of Lemma 1.2.6, showing that $P\{M_n \leqslant u_n\} \to e^{-\tau}$ if and only if $P\{\xi_1 > u_n\} \sim \tau/n$ as $n \to \infty$. This has been extended to m-dependent cases in [39] and strong mixing in [25], [27]. We shall see that the "if" part of this result remains true for stationary sequences satisfying both the conditions $D(u_n)$, $D'(u_n)$. This yields some important consequences, such as the fact that (under such conditions) a limiting distribution of $M_n = \max(\xi_1, \xi_2, \ldots, \xi_n)$ is the same as if the ξ's are i.i.d., and thus depends only on the marginal d.f. of the ξ's. It would also be of interest to determine whether the "only if" part of Lemma 1.2.6 is also true under $D(u_n)$, $D'(u_n)$ (a currently open question). It is known that this is true under further conditions, including strong mixing and a condition of $D'(u_n)$-type [27]. We shall comment further on this after Theorem 2.3.2.

Our main result is the following. We sketch the proof and refer the reader to [23] for full details if these are desired.

THEOREM 2.3.1: *Suppose that $D(u_n)$, $D'(u_n)$ (i.e., (2.1.4) and (2.1.6)) hold for the stationary sequence $\{\xi_n\}$, where the $\{u_n\}$ satisfy (writing $F = F_1$, the marginal* d.f. *of the ξ_i)*

$$1 - F(u_n) = \tau/n + o(1/n). \tag{2.3.1}$$

Then

$$P\{M_n \leqslant u_n\} \to e^{-\tau} \qquad \textit{as } n \to \infty. \tag{2.3.2}$$

Sketch of proof: Fix an integer k and write $N = nk$. We have, clearly,

$$P\{M_n \leqslant u_N\} = 1 - P\left(\bigcup_{j=1}^{n} \{\xi_j > u_N\}\right) \geqslant 1 - nP\{\xi_1 > u_N\}$$

so that (2.3.1) gives

$$\liminf P\{M_n \leqslant u_N\} \geqslant 1 - \tau/k. \tag{2.3.3}$$

By a similar calculation we have (using stationarity) that

$$P\{M_n \leqslant u_N\} \leqslant 1 - nP\{\xi_1 > u_N\} + n \sum_{j=2}^{n} P\{\xi_1 > u_N, \xi_j > u_N\}$$

so that by (2.3.1) and $D'(u_n)$,

$$\limsup_n P\{M_n \leqslant u_N\} \leqslant 1 - \tau/k + o(1/k).$$

Hence by (2.3.3) and Lemma 2.2.4 we obtain

$$(1 - \tau/k)^k \leqslant \liminf P\{M_N \leqslant u_N\} \leqslant \limsup P\{M_N \leqslant u_N\}$$
$$\leqslant (1 - \tau/k + o(1/k))^k. \tag{2.3.4}$$

To complete the proof it may be shown that N may be replaced by n and then the result follows at once by letting $k \to \infty$. To do this we choose r (depending on n), such that $rk \leqslant n < (r + 1)k$ and note that

$$P\{M_n \leqslant u_n\} \geqslant P\{M_{(r+1)k} \leqslant u_{(r+1)k}\} - P\{u_n < M_{(r+1)k} \leqslant u_{(r+1)k}\}$$

from which it follows by some calculation that

$$\liminf P\{M_n \leqslant u_n\} \geqslant \liminf P\{M_{(r+1)k} \leqslant u_{(r+1)k}\} \geqslant (1 - \tau/k)^k.$$

A similar calculation yields the right-hand inequality of (2.3.4).

Write now $\hat{M}_n$ for the maximum of n i.i.d. random variables with the same marginal d.f. F_1 as the ξ_i (following [25] we may call these the "independent sequence associated with the ξ_i"). Then we have the following immediate corollary of the theorem.

COROLLARY: *Suppose that $D(u_n)$, $D'(u_n)$ are satisfied for the stationary sequence $\{\xi_n\}$ and that $P\{\hat{M}_n \leqslant u_n\} \to \theta > 0$. Then also $P\{M_n \leqslant u_n\} \to \theta$.*

Proof: By Lemma 1.2.6, $\{1 - F_1(u_n)\} = \tau/n + o(1/n)$ with $\tau = -\log \theta$, so that (2.3.1) holds and hence so does (2.3.2), which is the desired result.

As a further immediate corollary we have the following result which shows that under conditions of the type $D(u_n)$, $D'(u_n)$ any limiting distribution for the maximum $\hat{M}_n$ of the associated independent sequence is inherited by that for the stationary sequence.

THEOREM 2.3.2: *With the established notation, suppose that*

$$P\{a_n(\hat{M}_n - b_n) \leqslant x\} \to G(x)$$

for some constants $a_n > 0$, b_n and some nondegenerate (Type I, II,

III) d.f. *G*. *Suppose that* $D(u_n)$, $D'(u_n)$ *are satisfied when* $u_n = x/a_n + b_n$ *for each* x. *Then*

$$P\{a_n(M_n - b_n) \leqslant x\} \to G(x).$$

Proof: This is immediate from the previous corollary when $G(x) > 0$ and is readily extended by continuity of G to points x where $G(x) = 0$.

This theorem shows that (under certain conditions at least), if $\hat{M}_n$ has some nondegenerate asymptotic distribution, then M_n has the same one (with the same normalizing constants). It does not preclude the possibility that M_n (normalized) might have a nondegenerate limiting distribution (necessarily of extreme value type by Theorem 2.2.1) whereas $\hat{M}_n$ has none, however it is normalized. It would of course be of interest to know (by proof or counterexample) whether or not this can occur. It is known that it cannot occur under certain more stringent conditions (including strong mixing—cf. remark at the start of this section, and [27]), and we suspect that the same holds here.

2.4. Point processes, and the Poisson character of high level exceedances. It turns out that a number of interesting results may be obtained by looking at the exceedances (by $\{\xi_n\}$) of a high level u (i.e., the points i such that $\xi_i > u$—cf. Chap. I) as a stochastic point process.

By a *point process* (on the real line) we mean simply a series of events occurring in time (or some other parameter) according to some probabilistic law, simple examples being the occurrence of telephone calls, radioactive disintegrations, and so on. We shall not need to involve ourselves with the detailed probabilistic structure of point process theory, but merely use one very simple but powerful theorem from this theory. (An excellent account of structural theory of point processes, within the framework of *random measures*, is given in [20] for the reader interested in a general and rigorous treatment.)

We shall write $N(B)$ for the number of events of a point process in the (Borel) set B. (Usually the set B of interest will be a (semi-closed) interval $(a, b]$ or finite union of such intervals.) One of the simplest and most useful point processes is the *Poisson Process*, which is defined by the requirement that $N(B)$ have a Poisson distribution with mean $\lambda m(B)$ ($\lambda > 0$, and $m(B)$ is the Lebesgue measure of B), for each B, and that $N(B_1), \ldots, N(B_k)$ be mutually independent whenever $B_1, \ldots, B_k$ are disjoint sets. Thus for example the probability of r events in the interval $(a, b]$ is

$$P\{N(a, b]) = r\} = e^{-\lambda(b-a)}[\lambda(b - a)]^r/r! \qquad (r = 0, 1, 2, \ldots)$$

and the probability of $r_1, r_2, \ldots, r_k$ events in each of the *disjoint* intervals $(a_1, b_1], \ldots, (a_k, b_k]$ is just the product of the corresponding Poisson probabilities $P\{N(a_i, b_i]) = r_i\}$. The parameter λ is usually called the intensity of the Poisson process and we have $\lambda = EN((0, 1])$ or more generally $EN(B) = \lambda m(B)$ for any (Borel) set B.

The exceedances of a level by a stochastic sequence $\{\xi_n\}$ do not usually form a Poisson process. Indeed, even when the ξ_j are i.i.d., we have, as in Section 1.3, a binomial distribution for the number of exceedances in a given "time" interval. However, in that section we used the convergence of the binomial to the Poisson distribution to obtain the approximate Poisson character of the number of exceedances when the level (u_n) became high. This suggests that we might expect the exceedances of a high level to exhibit, at least approximately, some of the properties of a Poisson process when the $\{\xi_n\}$ are i.i.d. and, we might hope, also when $\{\xi_n\}$ is a stationary sequence and conditions $D(u_n)$, $D'(u_n)$ hold. As we shall see, this turns out to be the case.

To explore this fully we shall need to talk about convergence of a sequence of point processes ($\{N_n\}$, say) to a point process N, *in distribution* as $n \to \infty$ (to be denoted by $N_n \xrightarrow{d} N$). We shall not need to define this in detail, but merely note that it implies, in particular, the convergence of the joint distribution of $(N_n(B_1), N_n(B_2), \ldots, N_n(B_k))$ to that of $(N(B_1), N(B_2), \ldots, N(B_k))$ as $n \to \infty$, for any choice of k and all $B_1, B_2, \ldots, B_k$ whose boundaries have zero Lebesgue measure. A full discussion of such "weak convergence of point processes" may be found in [20], but we need only the following very simple but useful result [20, Theorem 4.7]. In this we shall call a point process *simple* if there is zero probability of the occurrence of multiple events (i.e., of two or more events occurring simultaneously).

THEOREM 2.4.1: *Let $\{N_n\}$ be a sequence of point processes, and let N be a simple point process such that, for any fixed a, $P\{N(\{a\}) = 0\} = 1$. Suppose that*

(i) $P\{N_n(B) = 0\} \to P\{N(B) = 0\}$ *for all sets B of the form* $\bigcup_1^r (a_i, b_i]$ $(a_1 < b_1 < a_2 \cdots < a_r < b_r)$

(ii) $\limsup_{n\to\infty} EN_n\{(a, b]\} \leqslant EN\{(a, b]\}$ *for all finite $a < b$. Then* $N_n \xrightarrow{d} N$.

From this result we see that convergence of N_n to N in distribution can be deduced simply from convergence of $P\{N_n(B) = 0\}$ to $P\{N(B) = 0\}$ for sets $B = \bigcup_1^r (a_i, b_i]$, along with (ii) which holds certainly if $EN_n\{(a, b]\} \to E(N(a, b])$.

It is quite easy to apply this result now to the point processes consisting of the exceedances of a sequence of levels $\{u_n\}$ by a stationary sequence $\{\xi_n\}$ satisfying $D(u_n)$, $D'(u_n)$, and (2.3.1). To do this we must first make a simple change of time scale (by the factor $1/n$). The exceedances of the level u_n form a point process with events restricted to (a subset of) the integers $1, 2, 3, \ldots$. The events of our point process N_n will occur at (a subset of) points $1/n, 2/n, 3/n, \ldots$, an event occurring at i/n if $\xi_i > u_n$. That is $N_n\{(a, b]\}$ is the number of exceedances of u_n by ξ_i in $a < i/n \leqslant b$, i.e., in $(na, nb]$.

THEOREM 2.4.2: *Let $\{\xi_n\}$ be a stationary sequence and let $u_n = u_n(\tau)$ satisfy* (2.3.1). *Suppose that $D(u_n)$ and $D'(u_n)$ are satisfied for each $\tau > 0$. Then $N_n \xrightarrow{d} N$ where N_n is the point process of exceedances of u_n (with time scale change) and N is a Poisson process with intensity τ.*

Proof: We need to verify Conditions (i) and (ii) of Theorem 2.4.1. (ii) is simple, for we may write $N_n(a, b] = \sum_{a < j/n \leqslant b} \chi_j$ where $\chi_j = 1$

if $\xi_j > u_n$ and $\chi_j = 0$ otherwise. Hence (using "[ ]" to indicate integer part),

$$EN_n(a, b] = \sum_{an<j\leq bn} E\chi_j = \sum_{an<j\leq bn} P\{\chi_j = 1\}$$

$$= ([nb] - [na])(1 - F_1(u_n)) \sim n(b - a)(1 - F_1(u_n))$$

$$\to \tau(b - a)$$

by (2.3.1). But this is just $EN\{(a, b]\}$, so that (ii) follows.

We shall sketch the proof of (i) (again full details appear in [23]). The key step is to note first that if $0 < a < b$

$$P\{N_n((a, b]) = 0\} = P\{M((an, bn]) \leqslant u_n\} = P\{M_{[bn]-[an]} \leqslant u_n\}$$

by stationarity. Now from Theorem 2.3.1 we may show (by some obvious calculation) that for any $\alpha > 0$

$$P\{M_{[\alpha n]} \leqslant u_n(\tau)\} \to e^{-\alpha\tau} \qquad \text{as } n \to \infty.$$

By applying this result to $\alpha = b - a,\ b - a + h\ (h > 0)$, noting that $M_{[(b-a)n]} \leqslant M_{[bn]-[an]} \leqslant M_{[(b-a+h)n]}$, and letting $h \to 0$ it may be thus seen that

$$P\{N_n((a, b]) = 0\} \to e^{-\tau(b-a)}. \tag{2.4.1}$$

Finally let $B = \bigcup_1^r (a_i, b_i]$ where $a_1 < b_1 < a_2 \cdots < a_r < b_r$. Then if E_j is the set of integers $([na_j] + 1, [na_j] + 2, \ldots, [nb_j])$, we have

$$P\{N_n(B) = 0\} = P\left(\bigcap_{j=1}^{r} \{M(E_j) \leqslant u_n\}\right)$$

$$= \prod_{j=1}^{r} P\{N_n((a_j, b_j]) = 0\}$$

$$+ \left[P\left(\bigcap_{j=1}^{r} \{M(E_j) \leqslant u_n\}\right) - \prod_{j=1}^{r} P\{M(E_j) \leqslant u_n\}\right].$$

By (2.4.1) the first term converges to $\prod_{j=1}^r e^{-\tau(b_j - a_j)} = e^{-\tau m(B)}$ which is $P\{N(B) = 0\}$. On the other hand, by Lemma 2.1.1, the modulus of the remaining difference of terms is dominated by $(r - 1)\alpha_{n,[n\lambda]}$, where $\lambda = \min_{1 \leqslant j \leqslant r-1} (a_{j+1} - b_j)$. Since $\alpha_{n,l}$ is nonincreasing in l and eventually $[n\lambda] > l_n$ ($l_n = o(n)$—cf. (2.1.4)), we have $\alpha_{n,[n\lambda]} \to 0$ as $n \to \infty$ so that $P\{N_n(B) = 0\} \to P\{N(B) = 0\}$, as desired.

2.5. Distribution of kth largest values—fixed ranks. We shall now look briefly at the asymptotic properties of the distribution of $M_n^{(k)}$ first where k is fixed, and in the next section for $k = k_n \to \infty$.

Asymptotic results for fixed ranks k follow simply from the Poisson properties of the previous section. To see this we give first an immediate generalization of Lemma 1.3.2.

LEMMA 2.5.1: *Let $\{\xi_n\}$ be a stationary sequence and let $\{u_n\}$ satisfy* (2.3.1). *Let $D(u_n)$, $D'(u_n)$ hold and write S_n for the number of exceedances of u_n by ξ_i for $1 \leqslant i \leqslant n$. Then, for each k,*

$$P\{S_n \leqslant k\} \to e^{-\tau} \sum_{j=0}^{k} \tau^j/j!. \tag{2.5.1}$$

Proof: With the notation of the previous section, S_n is just $N_n((0, 1])$ which has a limiting Poisson distribution with mean τ, by Theorem 2.4.2.

The next result generalizes the first part of Theorem 1.3.3 to apply to our stationary sequences.

THEOREM 2.5.2: *Let $\{\xi_n\}$ be a stationary sequence, and suppose that*

$$P\{a_n(\hat{M}_n - b_n) \leqslant x\} \to G(x)$$

where $\hat{M}_n$ is the maximum of the first n terms of the independent sequence associated with $\{\xi_n\}$, and G is nondegenerate. If $D(u_n)$, $D'(u_n)$ hold for each u_n of the form $x/a_n + b_n$ $(-\infty < x < \infty)$, then

$$P\{a_n(M_n^{(k)} - b_n) \leqslant x\} \to G(x) \sum_{j=0}^{k-1} (-\log G(x))^j/j!. \tag{2.5.2}$$

Proof: The assumptions of the theorem imply (2.3.1) by Lemma 1.2.6 with $\tau = -\log G(x)$, $u_n = x/a_n + b_n$ and hence

$$P\{M_n^{(k)} \leqslant u_n\} = P\{S_n \leqslant k - 1\}$$

from which the result follows by Lemma 2.5.1.

This theorem shows that if the marginal d.f. of the stationary sequence $\{\xi_n\}$ belongs to the domain of attraction of the extreme value d.f. G, then $M_n^{(k)}$ is attracted to $G(x)\sum_{j=0}^{k-1}[-\log G(x)]^j/j!$ (provided $D(u_n)$, $D'(u_n)$ hold with the appropriate $\{u_n\}$). Thus provided $\hat{M}_n$ has *some* nondegenerate limiting distribution, a limiting distribution for any $M_n^{(k)}$ must be of the above form. Moreover its existence for some k then implies its existence, and its form, for all k (again if $D(u_n)$, $D'(u_n)$ hold). Of course it is potentially possible for $\hat{M}_n$ to have no nondegenerate limiting distribution, whereas some $M_n^{(k)}$ may have a nondegenerate limit. The question of whether this can occur or not is open (see also the comments after Theorem 2.3.2).

Finally we note that it is possible to consider exceedances of more than one level, and thus to obtain important *joint* asymptotic distributions (such as that of M_n, $M_n^{(k)}$). We do not, however, pursue this here.

2.6. Variable ranks. It is of course of interest to determine the extent to which the results of Chapter 1 for the k_nth largest of n i.i.d. random variables hold under weak mixing conditions. It seems intuitively reasonable to expect that at least some of these results might hold under conditions of $D(u_n)$, $D'(u_n)$ type, though the rates of convergence to zero might need to be rather faster. Since this work is at an early stage of development, we shall merely make some brief comments in indicating the current situation regarding these topics.

It turns out that for $k_n = n^\theta$ (where $\theta < \frac{2}{5}$), if the "mixing sequence" α_{n,l_n} of $D(u_n)$ tends to zero sufficiently fast (e.g., exponentially) and if a condition of $D'(u_n)$ type holds, then the direct part of the key result Lemma 1.4.2 holds. From this and the i.i.d. results one may see (as in the previous section for fixed ranks) that if the k_nth largest $\hat{M}_n^{(k_n)}$ of n i.i.d. random variables (with the same marginal d.f. as the ξ_i) has an asymptotic distribution, then $M_n^{(k_n)}$ has the same asymptotic distribution.

These results may be obtained by essentially the same methods as those used for the fixed ranks case—though the details are more involved. (A full discussion is to be found in [40].)

An alternative (and closely related) approach is also worthy of mention (and in fact complements the previous method by including the case $k_n = n^\theta$ for $\frac{2}{5} \leqslant \theta < 1$). If we write χ_j for the indicator r.v. which is unity if $\xi_j > u_n$ (as in Section 1.4), then we are interested in the asymptotic form of $P\{S_n \leqslant k_n\}$, where $S_n = \sum_1^n \chi_j$. This was discussed in Section 1.4, using some of the tools of ordinary central limit theory for (arrays of) i.i.d. random variables. On the other hand, appropriate central limit theory has been developed [12 and references therein] for r.v.'s satisfying strong mixing conditions. But since strong mixing for the r.v.'s $\{\chi_j\}$ is much weaker than strong mixing for the original sequence $\{\xi_j\}$, such assumptions are likely to be relevant for us.

By using this method, it is again possible to obtain conditions under which $M_n^{(k_n)}$ has the same nondegenerate limiting distribution as $\hat{M}_n^{(k_n)}$, the k_nth largest of the independent sequences associated with $\{\xi_j\}$. It is worth noting, incidentally, that the conditions of these results are satisfied by stationary normal sequences whose covariances r_n tend to zero sufficiently rapidly (e.g., exponentially) as $n \to \infty$. (The precise rate of convergence required depends on the sequence $\{k_n\}$.)

The results mentioned above refer primarily to the intermediate rank case—though similar results may be expected for central ranks. For a full discussion of the available results of this type, of relations between the mixing conditions, and of certain ("Bahadur") representations of the ordered values (cf. [1]), the reader is referred to [40].

2.7. Stationary normal sequences. In this section we shall show how the previous results apply to a stationary normal sequence $\{\xi_n\}$ under appropriate conditions on its covariance function.

To say that the stationary sequence $\{\xi_n\}$ is *normal* (or *Gaussian*) means that the joint distribution of $(\xi_{i_1}, \xi_{i_2}, \ldots, \xi_{i_k})$ is a k-dimensional normal distribution for any choice of k and $i_1, i_2, \ldots, i_k$. For simplicity we shall assume that $E\xi_n = 0$ and var $\xi_n = 1$ (these are in any case independent of n by stationarity). The covariance sequence $\text{cov}(\xi_1, \xi_{1+n})$ $(=\text{cov}(\xi_j, \xi_{j+n}) = E\xi_j\xi_{j+n}$ for all j) will be denoted by r_n. In particular we have $r_0 = E\xi_j^2 = 1$. The sequence $\{r_n\}$ determines the covariance matrix of any group $(\xi_{i_1}, \xi_{i_2}, \ldots, \xi_{i_k})$ and hence their joint (normal) distribution.

The aim is to show that $D(u_n)$ and $D'(u_n)$ hold if r_n satisfies suitable conditions—the simplest such sufficient condition being $r_n \log n \to 0$ as $n \to \infty$. This imposes very little restriction at all on the sequence and hence the limiting results depending on $D(u_n)$ and $D'(u_n)$ will be widely true for normal sequences. First it will be convenient to state a small technical lemma.

LEMMA 2.7.1: *Let* $\{u_n\}$ *satisfy* (2.3.1), *i.e., in this context*

$$1 - \Phi(u_n) \sim \tau/n \tag{2.7.1}$$

where Φ *denotes the standard normal* d.f. *Then*

$$n \sum_{j=1}^{n} |r_j| \exp\{-u_n^2/(1 + |r_j|)\} \to 0 \quad \text{as } n \to \infty \tag{2.7.2}$$

if

$$r_n \to 0 \quad \text{and} \quad \frac{1}{n} \sum_{j=1}^{n} |r_j| \log j \, e^{\gamma |r_j| \log j} \to 0 \quad \text{as } n \to \infty \tag{2.7.3}$$

for some $\gamma > 2$. *Further* (2.7.3) *holds if either* $r_n \log n \to 0$ *as* $n \to \infty$ *or if* $\sum |r_j|^p < \infty$ *for some* $p > 0$.

Method of Proof: To prove that (2.7.3) implies (2.7.2) the sum $\sum_{j=1}^{n}$ may be split into three parts—the first being $\sum_{j=1}^{[n^\alpha]}$ where $0 < \alpha < (1 - \delta)/(1 + \delta)$ ($\delta = \sup_{n \geq 1} |r_n| < 1$, since $r_n \to 0$, and if $|r_n| = 1$ for some $n \geqslant 1$, then $r_n = 1$ infinitely often). Some obvious estimation, using the fact that $\exp(-u_n^2/2) \sim K u_n/n$ shows that this part tends to zero. By similar calculations the remaining parts, viz., $\sum_{j=[n^\alpha]}^{[n^\beta]}$ and $\sum_{j=[n^\beta]}^{n}$ (where $\beta = [2/\gamma]$) may be appropriately dominated and thus shown to tend to zero. (Again full details may be found in [23].)

The heart of the proof that $D(u_n)$, $D'(u_n)$ hold under appropriate conditions on r_n is contained in the following lemma. We give the proof in some detail since it involves an instructive method developed in various ways by Slepian [35], Berman [4], [6], Cramér (see [10]).

LEMMA 2.7.2: *Let $\{\xi_n\}$ be the stationary normal sequence defined above (and assume its joint distributions are nondegenerate, for simplicity). Let $1 \leqslant l_1 \leqslant \cdots \leqslant l_s$ be integers. Then for any u*

$$\left|P\left(\bigcap_{j=1}^{s}\{\xi_{l_j} \leqslant u\}\right) - \Phi^s(u)\right| \leqslant K \sum_{1 \leq i < j \leq s} |\rho_{ij}| \exp\{-u^2/(1 + |\rho_{ij}|)\} \tag{2.7.4}$$

where $\rho_{ij} = r_{l_i - l_j}$ is the correlation between ξ_{l_i} and ξ_{l_j}, and K is a constant.

Proof: Let f_1 denote the joint (normal) p.d.f. of $\xi_{l_1}, \ldots, \xi_{l_s}$, based on the covariance matrix Λ_1 (with (i, j)th element ρ_{ij}). Then

$$P\left(\bigcap_{j=1}^{s}\{\xi_{l_j} \leqslant u\}\right) = \int_{-\infty}^{u} \cdots \int f_1(y_1, \ldots, y_s)\, dy_1 \ldots dy_s.$$

Let $\Lambda_0 = I$ (the identity matrix—which would have been the covariance matrix if the ξ_{l_j} were independent) and write $\Lambda_h = h\Lambda_1 + (1 - h)\Lambda_0$ $(0 \leqslant h \leqslant 1)$. Λ_h is positive definite, with units down the main diagonal, and elements $h\rho_{ij}$ for $i \neq j$. Let f_h be the s-dimensional normal density based on Λ_h and

$$F(h) = \int_{-\infty}^{u} \cdots \int f_h(y_1, \ldots, y_s)\, dy \qquad (dy = dy_1 \ldots dy_s).$$

The left-hand side of (2.7.4) is easily recognized as $|F(1) - F(0)|$. Now

$$|F(1) - F(0)| \leqslant \int_0^1 |F'(h)|\, dh \tag{2.7.5}$$

where

$$F'(h) = \int_{-\infty}^{u} \cdots \int_{-\infty}^{u} \frac{\partial f_h(y_1, \ldots, y_s)}{\partial h}\, dy.$$

The density f_h depends on h through the elements λ_{ij} of Λ_h (regarding f_h as a function of λ_{ij} for $i \leqslant j$, say). Thus

$$F'(h) = \sum_{i<j} \rho_{ij} \int_{-\infty}^{u} \cdots \int_{-\infty}^{u} \frac{\partial f_h}{\partial \lambda_{ij}}\, dy$$

since $\lambda_{ij} = h\rho_{ij}$ if $i < j$ and 1 if $i = j$. Now a useful property of the multidimensional normal density is that its derivative with respect to a covariance λ_{ij} is the same as the second mixed derivative with respect to the corresponding variables y_i, y_j (cf. [10, p. 26]). Thus

$$|F'(h)| \leqslant \sum_{i<j} |\rho_{ij}| \left| \int_{-\infty}^{u} \cdots \int_{-\infty}^{u} \frac{\partial^2 f_h}{\partial y_i \partial y_j}\, dy \right|.$$

The y_i and y_j integrations may be done at once to give

$$\sum_{i<j} |\rho_{ij}| \int_{-\infty}^{u} \cdots \int_{-\infty}^{u} f_h(y_i = y_j = u)\, dy',$$

where $f_h(y_i = y_j = u)$ denotes the function of $s - 2$ variables formed by putting $y_i = y_j = u$, the integration being over the remaining variables.

We can dominate the last integral by letting the remaining variables run from $-\infty$ to ∞. But $\int_{-\infty}^{\infty} \cdots \int f_h(y_i = y_j = u)\, dy'$ is just the bivariate density, evaluated at (u, u), of two standard normal random variables with correlation $h\rho_{ij}$, i.e.,

$$\frac{1}{2\pi(1 - h^2\rho_{ij}^2)^{1/2}}\, e^{-u^2/(1 + h\rho_{ij})}.$$

This gives

$$|F'(h)| \leqslant \frac{1}{2\pi} \sum_{i<j} \left|\rho_{ij}\right| (1 - \rho_{ij}^2)^{-1/2}\, e^{-u^2/(1 + |\rho_{ij}|)}, \qquad 0 \leqslant h \leqslant 1.$$

Since, by nondegeneracy, no $r_n = 1$ for $n \geqslant 1$, the r_n are actually uniformly bounded away from 1 (by a remark in the proof of Lemma 2.7.1), and thus $(1 - \rho_{ij}^2)^{-1/2}$ is bounded above. The conclusion of the lemma follows from (2.7.5).

It is easy now to obtain the main results *directly* from this lemma, e.g., we may take $s = n$, $l_j = j$, $u = u_n$ in (2.7.4) and, by using, e.g., Lemma 2.7.1 to show that the right-hand side tends to zero, see that $P\{M_n \leqslant u_n\}$ is approximated by $\Phi^n(u_n)$ which converges to $e^{-\tau}$, from the i.i.d. theory. However, essentially the same small calculation may be used to verify $D(u_n)$, $D'(u_n)$, as the following lemma shows, and hence we may appeal to the general theory of Section 2.3 for our results. Even though this is a less direct route in the special case of normality, it demonstrates the application of the theory, and indicates that the conditions $D(u_n)$, $D'(u_n)$ are really very weak indeed.

LEMMA 2.7.3: *Suppose that the covariances* $\{r_n\}$ *of the stationary normal sequence* $\{\xi_n\}$ *satisfy* (2.7.3) (*which will hold if, e.g.,* $r_n \log n \to 0$ *as* $n \to \infty$). *Then both* $D(u_n)$, $D'(u_n)$ *hold if* $1 - \Phi(u_n) \sim \tau/n$.

Proof: From (2.7.4), with $s = 2$, $l_1 = 1$, $l_2 = j$, $N = nk$ we have

$$|\Pr\{\xi_1 \leqslant u_N, \xi_j \leqslant u_N\} - \Phi^2(u_N)| \leqslant K|r_{j-1}|\, e^{-u_N^2/(1+|r_{j-1}|)}$$

whence, by simple manipulation,

$$|\Pr\{\xi_1 > u_N, \xi_j > u_N\} - (1 - \Phi(u_N))^2| \leqslant K|r_{j-1}|\, e^{-u_N^2/(1+|r_{j-1}|)}.$$

Thus

$$n \sum_{j=2}^{n} \Pr\{\xi_1 > u_N, \xi_j > u_N\} \leqslant \frac{\tau^2}{k^2} + KN \sum_{j=1}^{N} |r_j|\, e^{-u_N^2/(1+|r_j|)} + o(1)$$

from which $D'(u_n)$ follows by Lemma 2.7.1.

It follows also from (2.7.4) that if $1 \leqslant l_1 \leqslant \cdots \leqslant l_s \leqslant n$, then

$$|F_{l_1, \ldots, l_s}(u_n) - \Phi^s(u_n)| \leqslant Kn \sum_{j=1}^{n} |r_j|\, e^{-u_n^2/(1+|r_j|)}.$$

Suppose now that $1 \leqslant i_1 < \cdots < i_p < j_1 < \cdots < j_q \leqslant n$. Identifying $\{l_1, \ldots, l_s\}$ in turn with $\{i_1, \ldots, i_p, j_1, \ldots, j_q\}$, $\{i_1, \ldots, i_p\}$ and $\{j_1, \ldots, j_q\}$ we thus have

$$|F_{i_1, \ldots, i_p, j_1, \ldots, j_q}(u_n) - F_{i_1, \ldots, i_p}(u_n) F_{j_1, \ldots, j_q}(u_n)|$$

$$\leqslant 3Kn \sum_{j=1}^{n} |r_j|\, e^{-u_n^2/(1+|r_j|)}$$

which tends to zero by Lemma 2.7.1.

The main results now follow at once.

THEOREM 2.7.4: *If* $\{\xi_n\}$ *is a stationary normal sequence whose covariances* $\{r_n\}$ *satisfy either* (i) $r_n \log n \to 0$ *or* (ii) $\sum_n |r_n|^p < \infty$ *for some* $p > 0$ *or* (iii) *the weaker condition* (2.7.3), *then* $P\{M_n \leqslant u_n\} \to e^{-\tau}$ *if* $1 - \Phi(u_n) \sim \tau/n$. ($u_n$ *is then given in terms of* n, τ *by* (1.2.7).)

Proof: This follows at once from Lemmas 2.7.1, 2.7.3, and Theorem 2.3.1.

THEOREM 2.7.5: *Under any of the covariance conditions of Theorem* 2.7.4 *we have*

$$P\{a_n(M_n - b_n) \leqslant x\} \to \exp(-e^{-x}) \qquad \text{as } n \to \infty$$

where

$$a_n = (2 \log n)^{1/2},$$

$$b_n = (2 \log n)^{1/2} - \tfrac{1}{2}(2 \log n)^{-1/2}[\log \log n + \log 4\pi].$$

Proof: Since by Theorem 1.2.7 the result holds for the associated independent sequence, we have from Lemma 1.2.6 that (1.2.3) holds, i.e., $(1 - \Phi(u_n)) \sim \tau/n$ with $u_n = x/a_n + b_n$, $\tau = e^{-x}$. Hence by Lemma 2.7.3, $D(u_n)$, $D'(u_n)$ hold, so that the result follows from Theorem 2.3.2.

Finally, it follows from Theorem 2.5.2 that the kth largest $M_n^{(k)}$ of $\xi_1, \ldots, \xi_n$ has the asymptotic distribution (2.5.2), with $G(x) = \exp(-e^{-x})$.

3. STATIONARY PROCESSES IN CONTINUOUS TIME

3.1. General properties—Gnedenko's Theorem. Most of the properties to be considered in this chapter will be for stationary *normal* (Gaussian) processes, since that is the case which has received the most study and for which detailed results are known. The conditions to be used there will be direct analogues of the covariance conditions which, in the discrete case, implied the weak "distributional mixing" but were significantly weaker than the strong mixing condition.

We shall pursue these matters in the subsequent sections. In the present section we shall consider general properties related to extremes of stationary processes—not restricted to the normal case.

Throughout it will be assumed without comment that $\{\xi(t) : t \geqslant 0\}$ is a stationary process whose sample functions are, with probability one, everywhere continuous. By *stationarity* we mean that the joint distribution of $(\xi(t_1 + \tau), \xi(t_2 + \tau), \ldots, \xi(t_n + \tau))$ is independent of τ, for any choice of fixed $n, t_1, t_2, \ldots, t_n$. Continuity is a very weak assumption (see, e.g., [10], for sufficient conditions). It will be further assumed that the marginal d.f. F of each $\xi(t)$, is continuous, which will, of course, be true when we consider *normal* processes.

We shall be concerned with the maximum $M(T)$ of $\xi(t)$ in $0 \leqslant t \leqslant T$. By continuity $M(T)$ is (a.s.) an attained maximum and is a r.v., as the following sometimes useful lemma shows. (For example a statement such as "by stationarity $\max\{\xi(t) : a \leqslant t \leqslant a+T\}$ has the same distribution as $M(T)$" can be easily verified by this lemma.)

LEMMA 3.1.1: *Let* $D_n = \{r2^{-n}T : r$ *integer and* $0 \leqslant r \leqslant 2^n\}$ *and* $M_n = \max\{\xi(t) : t \in D_n\}$.
Then $M_n \uparrow M(T)$ a.s., *as* $n \to \infty$ *and* $M(T)$ *is a* r.v.

The proof of this is immediate.

The *upcrossings* of a level by $\xi(t)$ are closely related to its maximum and are useful for the discussion of extremal properties. Specifically, a continuous function $f(t)$ is said to have an *upcrossing* of the level u at t_0 if, for sufficiently small $\epsilon > 0$, $f(t) \leqslant u$ in $(t_0 - \epsilon, t_0)$ and $f(t) \geqslant u$ in $(t_0, t_0 + \epsilon)$ with strict inequality for at least one point in each interval. The number of upcrossings of the level u by $\xi(t)$ in $0 \leqslant t \leqslant T$ will be denoted by $N(T)$ or $N_u(T)$, which is a r.v. under the conditions we have assumed (see [10, Chap. 10] for details).

The events $\{M(T) \leqslant u\}$ and $\{N_u(T) = 0\}$ are closely related. For

$$P\{M(T) \leqslant u\} \leqslant P\{N_u(T) = 0\} \leqslant P\{M(T) \leqslant u\} + P\{\xi(0) \geqslant u\}, \tag{3.1.1}$$

so that if u is large and $P\{\xi(0) \geqslant u\}$ is small we may expect to approximate $P\{M(T) \leqslant u\}$ by $P\{N_u(T) = 0\}$.

The distributions of $M(T)$ and $N_u(T)$ are not usually exactly known. However, we may often obtain the moments of $N_u(T)$ in an explicit form, even though this becomes computationally difficult as the order increases. For example, under certain conditions on the bivariate distribution $p(x, z)$ of $\xi(t)$, and its derivative $\xi'(t)$ we have (cf. [10, p. 290])

$$EN_u(T) = T \int_{-\infty}^{\infty} |z| p(u, z)\, dz. \tag{3.1.2}$$

Even with this first moment we have the possibility of a useful bound for $P\{N_u(T) = 0\}$ and hence for $P\{M(T) \leqslant u\}$ given by

$$P\{N_u(T) = 0\} \geqslant 1 - EN_u(T).$$

The following small result gives a further connection between the upcrossings and the maximum.

THEOREM 3.1.2: *If the mean number of upcrossings $EN_u(T)$ of u by $\xi(t)$ in $0 \leqslant t \leqslant T$ is finite, and a continuous function of u at u_0, then the* d.f. *of $M(T)$ is continuous at u_0.*

The straightforward proof of this is given in [23].

It may be expected that general results for continuous time processes should hold under distributional mixing conditions along similar lines to those of the previous chapter for sequences. While our results in this direction are not sufficiently developed to describe here in any detail, it nevertheless is clear that such progress is possible. Certainly under the *strong* mixing condition (defined in a corresponding way to the sequence case of Section 2.1) general results of this kind are already apparent, such as the following version of Gnedenko's Theorem for continuous parameter processes.

THEOREM 3.1.3: *Assume the general conditions on the stationary process $\xi(t)$ stated at the beginning of this section, and also that $\xi(t)$ is strongly mixing. If*

$$P\{a_T(M_T - b_T) \leqslant x\} \to G(x) \qquad \textit{as } T \to \infty$$

for some constants $a_T > 0$, b_T, and if G is nondegenerate, then G has one of the three extreme value forms.

Proof: Let $\eta_n = \max\{\xi(t) : n - 1 \leqslant t \leqslant n\}$. Then $\{\eta_n\}$ is a strongly mixing stationary sequence, and if $M_n = \max(\eta_1, \eta_2, \ldots, \eta_n)$ we have

$$P\{a_n(M_n - b_n) \leqslant x\} = P\{a_n(M(n) - b_n) \leqslant x\} \to G(x)$$

so that the conditions of Theorem 2.2.1 hold, and thus G has one of the three extreme value forms.

It may well be possible to use the argument in the above result under weaker mixing conditions—such as an appropriate analog of $D(u_n)$—for continuous parameter processes. Further it may also be possible to obtain other results corresponding to those for sequences in this way. For example, suppose it is possible to calculate the tail of the distribution of the maximum of $\xi(t)$ in a fixed finite time interval (e.g., $0 \leqslant t \leqslant 1$), and that we have

$$P\{\max(\xi(t) : 0 \leqslant t \leqslant 1) > u\} \sim h(u) \qquad \text{as } u \to \infty,$$

say, where $h(u)$ is known. Then writing, as above, $\eta_j = \max(\xi(t) : j - 1 \leqslant t \leqslant j)$ we have again

$$M(n) = \max\{\xi(t) : 0 \leqslant t \leqslant n\} = \max\{\eta_1, \eta_2, \ldots, \eta_n\}.$$

Now if u_n may be chosen so that $h(u_n) \sim \tau/n$ as $n \to \infty$ and if $\{\eta_n\}$ satisfies $D(u_n)$ and $D'(u_n)$, then we have, by Theorem 2.3.1,

$$P\{M(n) \leqslant u_n\} \to e^{-\tau},$$

from which other distributional results may be obtained, as for the case of sequences. (Typically the same results will hold when the integer sequence n is replaced by a continuous $T \to \infty$.)

It is clear that some progress on questions of extreme values for general stationary processes is possible by these or related methods. However, we shall confine our attention in the subsequent sections of this chapter to the case of stationary *normal* processes, where many results are known. Our treatment will largely be along the lines of that suggested above, although we shall take a slightly more direct line than that of showing explicitly that the $D(u_n)$, $D'(u_n)$ conditions hold for the sequence $\{\eta_n\}$ defined above.

Finally in these general comments we mention the Poisson character of high level upcrossings. This will again be considered explicitly in the normal case, but it is important to realize that it is by no means tied to normality. In the case of sequences, the exceedances of a level u formed a point process, with Poisson character under certain wide conditions for large u values. In the continuous parameter case the upcrossings of the level u replace the exceedances in our consideration (since the exceedances are infinite in number when they occur). We shall show convergence of this point process to a Poisson process in the normal case in a similar way to that previously considered for sequences. But—as was the case for sequences—it is the dependence structure rather than normality which causes this property to hold.

3.2. Stationary normal processes—preliminary properties. From now on we shall assume that our stationary process $\xi(t)$ is also normal—that is, the joint distribution of $\xi(t_1), \xi(t_2), \ldots, \xi(t_n)$ is an n-dimensional normal distribution for any choice of $n, t_1, t_2, \ldots, t_n$. We shall assume that $\xi(t)$ has zero mean, and unit variance, i.e., $E\xi(t) = 0$, $E\xi^2(t) = 1$ for all t. The covariance $\operatorname{cov}(\xi(t), \xi(t+\tau)) = E\xi(t)\xi(t+\tau)$ is independent of t by stationarity and is the *covariance function*—denoted here by $r(\tau)$. $r(\tau)$ specifies the joint normal distribution of any $(\xi(t_1), \xi(t_2), \ldots, \xi(t_n))$—this being based on the covariance matrix whose (i, j)th element is $r(t_i - t_j)$.

For most of the discussion we shall be concerned with the "regular" case—that for which $r(\tau)$ is twice differentiable. In this case we may write

$$r(\tau) = 1 - \lambda_2\tau^2/2 + o(\tau^2) \tag{3.2.1}$$

where $\lambda_2 = -r''(0)$. Later we shall indicate the corresponding results for the more general case where we assume just

$$r(\tau) = 1 - C|\tau|^\alpha + o|\tau|^\alpha \qquad (0 < \alpha \leqslant 2), \tag{3.2.2}$$

which includes the important Ornstein-Uhlenbeck process (with $r(\tau) = e^{-\alpha|\tau|}$, i.e., $\alpha = 1$). The main difficulty which arises when

$\alpha < 2$ is that the paths are much less well behaved and may, for example, have infinitely many upcrossings of a level in a finite time, even though (3.2.2) is sufficient to guarantee continuous sample paths for any $0 < \alpha \leqslant 2$. Until further notice we shall assume that (3.2.1) holds (i.e., $\alpha = 2$).

As indicated already, the number $N_u(T)$ of upcrossings of the level u by the stationary normal process $\xi(t)$ in $(0, T)$, will play an important role. In this case the formula (3.1.2) reduces to

$$EN_u(T) = \frac{T}{2\pi} \lambda_2^{1/2} e^{-u^2/2}. \tag{3.2.3}$$

Equation (3.2.3) is proved directly in [10, Chap. 10]. The main idea of the proof is to choose a sequence of values $q = q_n \to 0$ (e.g., $q = 2^{-n}T$, as in Lemma 3.1.1) and to approximate $N_u(T)$ by the number of integers i for which $\xi((i-1)q) < u < \xi(iq)$ $(1 \leqslant iq \leqslant T)$. More precisely, if $N_u^{(q)}(T)$ denotes the number of points iq in the set $(0, T)$ such that $\xi((i-1)q) < u < \xi(iq)$, then it is readily checked that

$$EN_u(T) = \lim_{q\to 0} EN_u^{(q)}(T) = \lim_{q\to 0} \frac{T}{q} P\{\xi(0) < u < \xi(q)\}$$

which may be evaluated to give

$$\lim_{q\to 0} \frac{T}{q} P\{\xi(0) < u < \xi(q)\} = T \frac{\lambda_2^{1/2}}{2\pi} e^{-u^2/2} \tag{3.2.4}$$

(and hence (3.2.3) follows).

In the above calculation u is fixed. The following lemma gives a technically very useful result corresponding to (3.2.4) when we allow u to tend to infinity as $q \to 0$.

LEMMA 3.2.1: *Suppose* (3.2.1) *holds. If* $u \to \infty$ *as* $q \to 0$ *in such a way that* $qu \to a$ (*where* $0 \leqslant a < \infty$), *then*

$$\frac{1}{q} P\{\xi(0) < u < \xi(q)\} \sim \frac{\lambda_2^{1/2}}{2\pi} e^{-u^2/2}\nu_a \qquad \text{as } u \to \infty$$

where $\lim_{a\to 0} \nu_a = \nu_0 = 1$.

Sketch of Proof: It will be convenient to take $\lambda_2 = 1$ (which may be effected in any case by a change of time scale). Then if $\zeta = q^{-1}(\xi(q) - \xi(0))$ we have

$$P\{\xi(0) < u < \xi(q)\} = P\{u - q\zeta < \xi(0) < u\}$$
$$= \int_0^\infty dz \int_{u-qz}^{u} p(x, z)\, dx$$
$$= q \int_0^\infty dz \int_0^z p(u - qy, z)\, dy$$

where p is the joint (normal) p.d.f. of $(\xi(0), \zeta)$ and is based on the covariance matrix

$$\Sigma = \begin{bmatrix} 1 & q^{-1}(r(q) - 1) \\ q^{-1}(r(q) - 1) & 2q^{-2}(1 - r(q)) \end{bmatrix}$$

as is easily checked. By writing this density out we find that

$$e^{u^2/2} p(u - qy, z) = (2\pi|\Sigma|)^{-1/2} \exp\left\{\tfrac{1}{2}u^2\left[1 - \frac{2(1 - r(q))}{q^2|\Sigma|}\right]\right\}$$
$$\times \exp\left\{-\frac{1}{2|\Sigma|}\left[\frac{2}{q^2}(1 - r(q))(-2uqy + q^2y^2)\right.\right.$$
$$\left.\left. - \frac{2}{q}(r(q) - 1)(u - qy)z + z^2\right]\right\}.$$

Now from (3.2.1) (with $\lambda_2 = 1$), noting that $|\Sigma| \to 1$ and using the fact that $uq \to a$ we may readily show that

$$e^{u^2/2} p(u - qy, z) \to \frac{1}{2\pi} e^{-a^2/8}\, e^{-1/2(-2ay + az + z^2)}.$$

It may be easily seen also that $e^{u^2/2}p(u - qy, z) \leqslant K_1\, e^{-K_2 z^2}$ in the integration range, and hence dominated convergence yields

$$2\pi \lim_{q \to 0} e^{u^2/2} \int_0^\infty dz \int_0^z p(u - qy, z)\, dy$$
$$= e^{-a^2/8} \int_0^\infty dz \int_0^z e^{-1/2(-2ay + az + z^2)} dy$$

which evaluates simply giving $(2\pi)^{1/2}\cdot[\Phi(a/2)-\Phi(-a/2)]/a$ (Φ, as always, denoting the standard normal d.f.) and the results follow by writing ν_a for this last expression.

Our next result is a more general form of Lemma 2.7.2 and we will state it without proof—since the proof requires only small and obvious changes from that of Lemma 2.7.2.

LEMMA 3.2.2: *Let* $\xi_1,\ldots,\xi_s$ *be jointly normal* r.v.'s *with zero means, unit variances, and covariance matrix* Λ_{ij}, *and* $\xi_1^*,\ldots,\xi_s^*$ *similarly with covariance matrix* Λ_{ij}^*. *Write* $\rho_{ij}=\max(|\Lambda_{ij}|,|\Lambda_{ij}^*|)$. *Then, for any u,*

$$\left|P\left\{\bigcap_{j=1}^{s}(\xi_j\leqslant u)\right\}-P\left\{\bigcap_{j=1}^{s}(\xi_j^*\leqslant u)\right\}\right|$$
$$\leqslant\frac{1}{2\pi}\sum_{1\leq i<j\leq s}|\Lambda_{ij}-\Lambda_{ij}^*|(1-\rho_{ij}^2)^{-1/2}\,e^{-u^2/(1+\rho_{ij})}.$$

Note that Lemma 2.7.2 is essentially the special case of this result for which Λ_{ij}^* is the identity matrix—i.e., $\xi_1^*,\ldots,\xi_n^*$ are independent.

Finally, in this series of preliminary results, we consider a special but very useful stationary normal process. Let η, ζ be independent standard normal r.v.'s and write

$$\xi^*(t)=\eta\cos t+\zeta\sin t.$$

It is clear that ξ^* is a normal process with zero mean and unit variance. Direct calculation shows that

$$E\xi^*(t)\xi^*(t+\tau)=\cos\tau$$

which is independent of t, leading to strict stationarity of $\xi^*(t)$ in the sense previously defined. Thus $\xi^*(t)$ is a stationary normal process with zero mean, unit variance, and covariance function $r(\tau)=\cos\tau$. We also have

$$\xi^*(t)=A\cos(t-\phi)$$

where it may be readily checked that A, ϕ are independent r.v.'s, A having a Rayleigh distribution with density $A\,e^{-A^2/2}$ ($A\geqslant 0$), and ϕ being the uniform over $(0,2\pi)$.

The distribution of the maximum $M^*(T)$ for this process can be obtained geometrically. However, it is more instructive (and simpler) to use properties of upcrossings. It is clear that $\lambda_2 = 1$ for this process and, writing $N_u^*(T) = N^*$, we have

$$E(N^*) = \frac{T}{2\pi} e^{-u^2/2}.$$

Now if $T < \pi$,

$$\begin{aligned}\{M^*(T) > u\} &= \{\xi^*(0) > u\} \cup \{\xi^*(0) \leqslant u, N^* \geqslant 1\} \\ &= \{\xi^*(0) > u\} \cup (\{N^* \geqslant 1\} - \{N^* \geqslant 1, \xi^*(0) > u\}).\end{aligned}$$

But if $\xi^*(0) > u$, the first upcrossing of u occurs after $t = \pi$, and hence $\{N^* \geqslant 1, \xi^*(0) > u\}$ is empty. Thus

$$\begin{aligned}\Pr\{M^*(T) > u\} &= 1 - \Phi(u) + \Pr\{N^* \geqslant 1\} \\ &= 1 - \Phi(u) + E(N^*) \qquad \text{(since } N^* = 0 \text{ or } 1) \\ &= 1 - \Phi(u) + \frac{T}{2\pi} e^{-u^2/2}. \qquad (3.2.4)\end{aligned}$$

This can be turned around, of course, to read

$$\Pr\{M^*(T) \leqslant u\} = \Phi(u) - \frac{T}{2\pi} e^{-u^2/2}. \qquad (3.2.5)$$

As a matter of interest it follows also for this process from (3.2.4) that (writing $\phi(u) = (2\pi)^{-1/2} e^{-u^2/2}$),

$$\frac{\Pr\{M^*(T) > u\}}{T\phi(u)} \to \left(\frac{\lambda_2}{2\pi}\right)^{1/2} \quad \text{as } u \to \infty \qquad (3.2.6)$$

(since $1 - \Phi(u) \sim \phi(u)/u$ and $\lambda_2 = 1$). This limit in fact holds under much more general conditions, as we shall see in the next section.

Note finally that for any $\omega > 0$ with A and ϕ as above, $\xi^*(t) = A \cos(\omega t - \phi)$ is a normal stationary process with covariance function $\cos \omega\tau$. We have

$$\Pr\{M^*(T) \leqslant u\} = \Phi(u) - \frac{\omega T}{2\pi} e^{-u^2/2} \qquad (\omega T < \pi),$$

which may be shown directly or from the above by time scale change. From this it follows that (3.2.6) again holds.

3.3. Extrema of stationary normal processes. In this section we assume that $\xi(t)$ is a stationary normal process, standardized so that $E\xi(t) = 0$, $E\xi^2(t) = 1$, and having covariance function $r(\tau) = 1 - \lambda_2\tau^2/2 + o(\tau^2)$ (equation (3.2.1)). We investigate the asymptotic distribution of $M(T) = \max\{\xi(t) : 0 \leqslant t \leqslant T\}$.

The essence of the method is really the fact noted in 3.1 that $M(T)$ is, e.g., when $T = n$, the maximum of n r.v.'s $\eta_1, \eta_2, \ldots, \eta_n$ where in turn η_i is the maximum of $\xi(t)$ in the interval $[i - 1, i]$. Hence if we can choose u_n so that $P\{\eta_i > u_n\} \sim \tau/n$, we may be able to obtain the asymptotic value of $P\{M(n) \leqslant u_n\}$, and hence the asymptotic distribution of $M(T)$.

If we know the asymptotic form of $P\{M(I) > u\}$ $(M(I) = \max\{\xi(t) : t \in I\})$, as $u \to \infty$ for a fixed interval I (e.g., $[0, 1]$), we could expect to use this to find $\{u_n\}$ satisfying $P\{M(I) > u_n\} \sim \tau/n$ and hence proceed as above. Thus the heart of our task is to find the asymptotic form of $P\{M(I) > u\}$ for a fixed interval I, as $u \to \infty$, and this is done in Lemma 3.3.1 below. However as noted in 3.1, while we essentially follow the approach outlined above, we shall not do so slavishly in terms of details—so that these may look somewhat different from the outline suggested above. In particular we shall need to (essentially) approximate the maximum of $\xi(t)$ in an interval by its maximum at judiciously chosen sampled time points. Nevertheless, the essential approach is really that indicated above.

In the following, $N_u(h)$, $N_u(I)$ will as usual denote the number of upcrossings of u by $\xi(t)$ for $0 \leqslant t \leqslant h$, and $t \in I$ (an interval or other set), respectively. Similarly $N_u^{(q)}(I)$ will, for any $q > 0$, be the number of points $iq \in I$ for which $\xi((i - 1)q) < u < \xi(iq)$, $N_u^{(q)}(h) = N_u^{(q)}((0, h])$. That is $N^{(q)}$ counts upcrossings of u by the sequence $\xi(kq)$.

LEMMA 3.3.1: *With the above assumptions and notation, as* $q \to 0$, $u \to \infty$, $uq \to a \geqslant 0$:

(i) $EN^{(q)}(h) = \nu_a h\mu + o(\mu)$ *where*

$$\mu(=\mu(u)) = EN_u(1) = (\lambda_2^{1/2}/2\pi)\, e^{-u^2/2}$$

and $\lim_{a\to 0} \nu_a = \nu_0 = 1$.

(ii) *For any interval* I *of length* h,

$$0 \leqslant P\Big\{\bigcap_{jq\in I} (\xi(jq) \leqslant u)\Big\} - P\{M(I) \leqslant u\} \leqslant (1 - \nu_a)\mu h + o(\mu).$$

Proof: (i) follows immediately from Lemma 3.2.1 since

$$EN^{(q)}(h) = \left[\frac{h}{q}\right] P\{\xi(0) < u < \xi(q)\} \sim \frac{h}{q} P\{\xi(0) < u < \xi(q)\}.$$

(ii) is also simple since clearly $\{M(I) \leqslant u\} \subset \bigcap_{jq \in I} \{\xi(jq) \leqslant u\}$ and it is easy to check that their difference implies that either $\xi(\alpha) > u$ (where α is the left endpoint of I), or $N_u(I) \geqslant 1$ and $N_u^{(q)}(I) = 0$. But $P\{\xi(\alpha) > u\} = 1 - \Phi(u) \sim \phi(u)/u = o(\mu)$ and

$$\begin{aligned} P\{N_u(I) \geqslant 1, N_u^{(q)}(I) = 0\} &\leqslant P\{N_u(I) - N_u^{(q)}(I) \geqslant 1\} \\ &\leqslant E\{N_u(I) - N_u^{(q)}(I)\} = \mu h - \nu_a \mu h + o(\mu) \end{aligned}$$

by (i) (and stationarity), using also the fact that $N_u - N_u^{(q)}$ is non-negative.

In the next lemma we shall show that the events $\cap\{\xi(jq) \leqslant k\}$ differ very little in probability from the corresponding events with ξ replaced by the special (harmonic) process ξ^* of the last section.

LEMMA 3.3.2: *Let $\xi(t)$ be as above and let $\xi^*(t)$ be the special stationary normal process $A \cos(\omega t - \phi)$ defined in Section 3.2, with covariance $r^*(\tau) = \cos \omega\tau$. Then (writing $\omega = \lambda_2^{1/2}$) we may choose $h > 0$ so that*

$$\mu^{-1} \left| P\left\{ \bigcap_{0 < jq \leq h} (\xi(jq) \leqslant u) \right\} - P\left\{ \bigcap_{0 < jq \leq h} (\xi^*(jq) \leqslant u) \right\} \right| \to 0$$

as $q \to 0$, $uq \to a > 0$.

Proof: Clearly $r^*(t) = 1 - \lambda_2 t^2/2 + o(t^2)$ so that $\psi(t) = r(t) - r^*(t) = o(t^2)$. Also if $\rho(t) = \max(|r(t)|, |r^*(t)|)$ then we have $\rho(t) = 1 - \lambda_2 t^2/2 + o(t^2)$. Now if there are s points jq in $(0, h]$ write $\xi_1, \ldots, \xi_s$ for $\xi(q), \ldots, \xi(sq)$, and $\xi_1^*, \ldots, \xi_s^*$ for $\xi^*(q), \ldots, \xi^*(sq)$. In the notation of Lemma 3.2.2 we then have writing $jq = t_j$,

$$\Lambda_{ij} = r(t_j - t_i), \qquad \Lambda_{ij}^* = r^*(t_j - t_i), \qquad \rho_{ij} = \rho(t_j - t_i)$$

and by that lemma,

$$\begin{aligned} &\left| P\left\{ \bigcap_{j=1}^{s} (\xi_j \leqslant u) \right\} - P\left\{ \bigcap_{j=1}^{s} (\xi_j^* \leqslant u) \right\} \right| \\ &\qquad\qquad \leqslant K \sum_{1 \leq i < j \leq s} \frac{\psi(t_i - t_j)}{|t_i - t_j|} e^{-u^2/(1+\rho_{ij})} \end{aligned}$$

where K is a constant (whose value may change from line to line) and use has been made of the fact that h may be chosen so that $(1-\rho^2(t))^{-1/2} \leqslant \text{const.}\,|t|^{-1}$ when $|t| \leqslant h$. Some obvious estimation, writing $\theta(t) = \psi(t)/t^2$ (and noting $qs \leqslant h$) now yields

$$\left|P\left\{\bigcap_{j=1}^{s}(\xi_j \leqslant u)\right\} - P\left\{\bigcap_{j=1}^{s}(\xi_j^* \leqslant u)\right\}\right|$$

$$\leqslant Ks\, e^{-u^2/2} \sum_{j=1}^{s} jq\theta(jq)\, e^{-(u^2/4)(1-\rho(jq))}. \tag{3.3.1}$$

The right-hand side of (3.3.1) is dominated by

$$K e^{-u^2/2} \sum_{j=1}^{s} \theta(jq) j\, e^{-\alpha j^2 q^2 u^2} \leqslant K e^{-u^2/2} \sum_{j=1}^{s} j\theta(jq)\eta^j$$

where α is a positive constant and η is constant, $0 < \eta < 1$, use being made here of the convergence $uq \to a > 0$. Now $s \sim h/q \to \infty$ and the terms of the sum $\sum_{j=1}^{s} j\theta(jq)\eta^j$ tend to zero as $q \to 0$. But $\theta(jq)$ is bounded (independently of (j, q)) for $jq \leqslant h$ and $\sum j\eta^j$ converges, so that $\sum_{j=1}^{s} j\theta(jq)\eta^j \to 0$ as $q \to 0$. This completes the proof.

The main result we will need is the following, giving the form of the tail of the distribution of $M(h)$ for h fixed. We shall simply show in this result that it is the same as that calculated already for the special process ξ^* used above.

LEMMA 3.3.3: *With the standard notation and assumptions about $\xi(t)$, we can find $h > 0$ such that*

$$P\{M(h) > u\} \sim \mu h \qquad \textit{as } u \to \infty. \tag{3.3.2}$$

Proof: We use the notation of the previous lemmas, and write $M^*N_u^*$, etc., for quantities based on the special process ξ^*, corresponding to M, N_u, etc.

If $a > 0$ we may apply Lemma 3.3.1 (ii) to both ξ and ξ^*, and combine this with Lemma 3.3.2 to obtain

$$\mu^{-1}|P\{M(h) \leqslant u\} - P\{M^*(h) \leqslant u\}| \leqslant 2(1-\nu_a)h + o(1) + \gamma_{q,u}$$

where $\gamma_{q,u} \to 0$ as $q \to 0$, $u \to \infty$, $qu \to a$. Thus, choosing $q = a/u$, letting $u \to \infty$ and finally $a \to 0$ we obtain

$$\mu^{-1}|P\{M(h) \leqslant u\} - P\{M^*(h) \leqslant u\}| \to 0 \qquad \text{as } u \to \infty.$$

Hence the same result holds with $>$ replacing $\leqslant$ in both places. But (3.3.2) holds for M^* by (3.2.6), with h replacing T. Hence (3.3.2) holds for M.

We have thus obtained the form of the tail of the distribution of the maximum over a small fixed interval. As noted previously, this is akin to the use of the asymptotic form of the tail distribution of each term in considering sequences. The remainder of this section will be devoted to obtaining the desired asymptotic distribution of $M(T)$, based on the result just obtained. We shall omit some of the details which are similar to those already given in full.

Associated with each T we define a level u such that the mean number μ of upcrossings of u by $\xi(t)$ per unit time satisfies $\mu T \to \tau$ where τ is a fixed constant. u is related to T in a similar way to that in which u_n is related to n, for sequences, and its precise form will be given later.

Let h be a positive constant as in Lemma 3.3.3—i.e., such that $P\{M(h) > u\} \sim \mu h$, and write $n = [T/h]$, the integer part of T/h. We then have the following relationships:

$$\mu T \to \tau \qquad n = [T/h] \sim T/h \sim \tau/\mu h \tag{3.3.3}$$

$$P\{M(h) > u\} \sim \mu h \sim \tau/n. \tag{3.3.4}$$

Now the interval $[0, nh]$ may be split into n intervals $[(i-1)h, ih]$. Choose $\epsilon < h$ and write $I_j = [(j-1)h, jh - \epsilon]$, $I_j^* = [jh - \epsilon, jh]$. Then the interval $[0, nh]$ is the union of the $2n$ intervals $I_1, I_1^*, \ldots, I_n, I_n^*$ which alternately have lengths $h - \epsilon, \epsilon$:

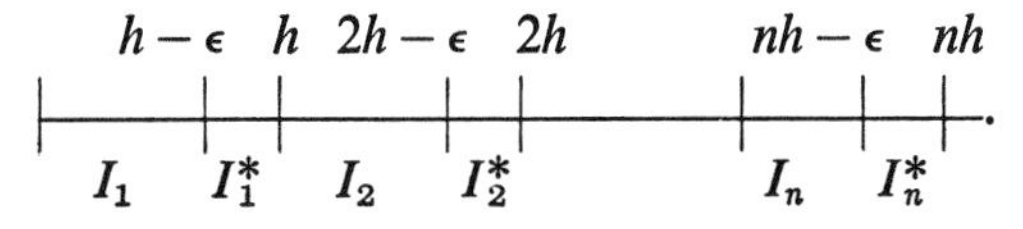

Finally choose $q \to 0$ as $n \to \infty$ (i.e., $T \to \infty$) such that $qu \to 0$. For convenience choose q so that h is always a multiple of q—so that the same number of points jq will lie in each of $I_1, I_2, \ldots$, and in each of $I_1^*, I_2^*, \ldots$.

In the following lemmas we proceed as follows:

(i) Approximate $P\{M(nh) \leqslant u\}$ by $P\{M(\bigcup_{k=1}^n I_k) \leqslant u\}$.

(ii) Approximate $P\{M(\bigcup_{k=1}^n I_k) \leqslant u\}$ by $P\{\bigcap \xi(jq) \leqslant u : jq \in \bigcup_{k=1}^n I_k\}$.

(iii) Approximate the latter probability by $P^n\{\bigcap_{jq\in I_1}(\xi(jq) \leqslant u)\}$.
(iv) Approximate this by $P^n\{M(I_1) \leqslant u\}$ and finally by $P^n\{M(h) \leqslant u\}$.
(v) Hence by taking limits show that $P\{M(nh) \leqslant u\} \to e^{-\tau}$ leading to $P\{M(T) \leqslant u\} \to e^{-\tau}$.

Specifically we have, then, the following lemmas.

LEMMA 3.3.4: *Under the standard conditions, and above notation (K constant independent of ϵ),*
(i) $0 \leqslant P\{M(\bigcup_{k=1}^{n} I_k) \leqslant u\} - P\{M(nh) \leqslant u\} \leqslant K\epsilon$
(ii) $0 \leqslant P\{\bigcap (\xi(jq) \leqslant u : jq \in \bigcup_{k=1}^{n} I_k)\} - P\{M(\bigcup_{k=1}^{n} I_k) \leqslant u\} \to 0.$

Proof: (i) is immediate since the difference of the probabilities is clearly nonnegative and dominated by $P\{\bigcup_1^n (M(I_k^*) > u)\} \leqslant nP\{M(I_1^*) > u\} \sim n\mu\epsilon$ by Lemma 3.3.3 and $n\mu\epsilon \to (\tau/h)\epsilon$ by (3.3.3).

(ii) Again the difference of the probabilities is clearly nonnegative, and is given by

$$P\left\{\bigcup_{k=1}^{n}\left(\bigcap_{jq\in I_k}\{\xi(jq) \leqslant u\} - \{M(I_k) \leqslant u\}\right)\right\}$$

$$\leqslant nP\left\{\bigcap_{jq\in I_1}(\xi(jq) \leqslant u) - (M(I_1) \leqslant u)\right\} \leqslant n\, o(\mu)$$

by Lemma 3.3.1 (ii), with $a = 0$. But $n\mu$ converges to a constant by (3.3.3) and hence $n\, o(\mu) \to 0$ as required.

To obtain the next approximation we need a condition akin to $D(u_n)$ or, more specifically, analogous to (2.7.2). The appropriate modification to (2.7.2) is contained in the following lemma, and we shall later see that it holds under very simple conditions indeed on r (as was also the case for sequences).

LEMMA 3.3.5: *Let $r(t) \to 0$ as $t \to \infty$. Suppose that as $T \to \infty$, $qu \to 0$, we have also*

$$\frac{T}{q} \sum_{\epsilon \leq jq \leq T} |r(jq)|\, e^{-u^2/(1+|r(jq)|)} \to 0 \tag{3.3.5}$$

for each fixed $\epsilon > 0$. Then
(i) $P\{\bigcap \xi(jq) \leqslant u : jq \in \bigcup_{k=1}^{n} I_k\} - P^n\{\bigcap_{jq\in I_1}(\xi(jq) \leqslant u)\} \to 0$
(ii) $P^n\{\bigcap_{jq\in I_1}(\xi(jq) \leqslant u)\} - P^n(M(I_1) \leqslant u) \to 0.$

Proof: (ii) is almost immediate (and does not require (3.3.5)) so we dispose of it first. If $y = P\{\bigcap_{jq\in I_1} (\xi(jq) \leqslant u)\}$, $x = P\{M(I_1) \leqslant u\}$, then by Lemma 3.3.1 (ii) ($a = 0$) we have $0 \leqslant x, y \leqslant 1$ and $y - x = o(\mu)$, and hence $y^n - x^n = (y - x)(y^{n-1} + y^{n-1}x + \cdots + x^{n-1}) \leqslant n(y - x) = o(n\mu) = o(1)$ by (3.3.3).

To prove (i), we use Lemma 3.2.2 again, with $\xi_j = \xi(jq)$ based on the covariance matrix Λ, $\Lambda_{ij} = r((i-j)q)$, and ξ_j^* based on the covariance matrix Λ^* where $\Lambda_{ij}^* = \Lambda_{ij}$ if iq and jq belong to the same interval I_k but is zero otherwise. That is, the group $\{\xi_j^* : jq \in I_k\}$ has the same joint (normal) distribution as $\{\xi_j : jq \in I_k)$ but the former groups are independent for different k-values. By Lemma 3.2.2 we have

$$|P\{\bigcap (\xi_j \leqslant u) : jq \in \bigcup I_k\} - P\{\bigcap (\xi_j^* \leqslant u) : jq \in \bigcup I_k\}|$$
$$\leqslant \frac{1}{2\pi} \Sigma' \, |r_{ij}|(1 - r_{ij}^2)^{-1/2} \, e^{-u^2/(1+|r_{ij}|)} \qquad (3.3.6)$$

in which $r_{ij} = r((i-j)q)$ and $\sum'$ denotes summation over all $i < j$ such that iq, jq belong to *different* intervals I_k. Now since $r(t) \to 0$ as $t \to \infty$ we may see that $|r(t)|$ is bounded away from 1 in $|t| \geqslant \epsilon$ and hence $(1 - r_{ij}^2)^{-1/2}$ is bounded above for all i, j in the sum considered (and fixed ϵ). Noting that there are certainly less than T/q terms with any given fixed value of $j - i$ in the sum we see that the right-hand side of (3.3.6) is, for fixed $\epsilon > 0$, dominated by the expression in (3.3.5), so that (i) follows.

These lemmas may now be simply combined to give the key result for $M(T)$. Before doing so we note the following simple sufficient condition for (3.3.5) to hold.

LEMMA 3.3.6: *If* $r(t) \log t \to 0$ *as* $t \to \infty$, *then* (3.3.5) *holds as long as* $qu \to 0$ *at an appropriately slow rate.*

The proof of this may be accomplished by splitting the sum into two parts—for $\epsilon \leqslant jq \leqslant T^\beta$ and $T^\beta \leqslant jq \leqslant T$, where β is chosen so that $0 < \beta < (1 - \delta)/(1 + \delta)$, δ being $\sup\{|r(t)| : |t| \geqslant \epsilon\}$ (<1). The details of this are quite straightforward and may be found in [5] or [23].

THEOREM 3.3.7: *Let the stationary normal process satisfy* (3.2.1) *and suppose that also* $r(t) \log t \to 0$ *as* $t \to \infty$. *Then if* u *is chosen so that the mean number* μ *of upcrossings of* u *per unit time* ($\mu = (\lambda_2^{1/2}/2\pi)\, e^{-u^2/2}$) *satisfies* $\mu T \to \tau$ *as* $T \to \infty$ *then*

$$P\{M(T) \leqslant u\} \to e^{-\tau} \qquad \text{as } T \to \infty. \tag{3.3.7}$$

In fact this is true if the condition $r(t) \log t \to 0$ is replaced by the even weaker (but more complicated) condition (3.3.5).

Proof: By combining Lemmas 3.3.4 and 3.3.5 (and using their notation) we have for fixed $\epsilon > 0$, if (3.3.5) holds,

$$\limsup_{n\to\infty} |P\{M(nh) \leqslant u\} - P^n\{M(I_1) \leqslant u\}| \leqslant K\epsilon$$

and hence, since ϵ is arbitrary,

$$P\{M(nh) \leqslant u\} - P^n\{M(I_1) \leqslant u\} \to 0. \tag{3.3.8}$$

Now $P\{M(I_1) > u\} \sim \mu h \sim \tau/n$ by Lemma 3.3.3 and (3.3.3) so that

$$P^n\{M(I_1) \leqslant u\} = (1 - \tau/n + o(1/n))^n \to e^{-\tau}.$$

Hence it follows from (3.3.8) that $P\{M(nh) \leqslant u\} \to e^{-\tau}$. $n = [T/h]$, and the same argument clearly holds for $n + 1$, so that $P\{M((n + 1)h) \leqslant u\} \to e^{-\tau}$. But $M(nh) \leqslant M(T) \leqslant M((n + 1)h)$ so that $P\{M(T) \leqslant u\} \to e^{-\tau}$, as required.

As a corollary we may at once obtain the double exponential limiting distribution for $M(T)$.

THEOREM 3.3.8: *Under the conditions* (3.2.1), *and* $r(t) \log t \to 0$, *the maximum* $M(T)$ *of the stationary normal process* $\xi(t)$ *in* $0 \leqslant t \leqslant T$ *satisfies*

$$P\{a_T(M(T) - b_T) \leqslant x\} \to \exp(-e^{-x})$$

where

$$\begin{cases} a_T = (2 \log T)^{1/2} \\ b_T = (2 \log T)^{1/2} + \log(\lambda_2^{1/2}/2\pi)/(2 \log T)^{1/2}. \end{cases} \tag{3.3.9}$$

Proof: The calculation here is simpler than that for the corresponding discrete problem. Specifically to apply Theorem 3.3.7 we must choose u so that $(\lambda_2^{1/2}/2\pi)e^{-u^2/2}T \to \tau$. By taking logarithms and writing $x = -\log \tau$ it is readily checked that we require

$$u = \frac{x}{a_T} + b_T + o(a_T^{-1})$$

and conversely that for such u, $\mu T \to \tau$. Hence

$$P\{M(T) \leqslant x/a_T + b_T\} \to e^{-\tau} = \exp(-e^{-x}),$$

as required.

3.4. Poisson nature of the upcrossings. It is again of interest to regard the upcrossings of a level u as a point process and to apply the weak convergence result Theorem 2.4.1. This will show that the upcrossings (after time scale change) converge weakly to a Poisson process as did the exceedances in the sequence case. In the exceedance case for sequences this led to the asymptotic distribution of the kth largest value. Correspondingly in this case we may obtain the asymptotic distribution of the kth largest local maximum.

It should again be mentioned that—although we shall not do so here—one may simultaneously consider the point processes consisting of upcrossings of more than one level. This leads to joint distributions of interest such as that of the kth and lth largest local maxima, or of the maxima in two different time intervals, or of the maximum and its location, and so on. We refer to [24] or [23] for details of such calculations.

Let us consider, then, the upcrossings of a high level u by the stationary normal process $\xi(t)$. As before μ will denote the mean number of such upcrossings per unit time. For each u we define a point process consisting of the upcrossings of u by the process

$$\xi_u^*(t) = \xi\left(\frac{\tau}{\mu}t\right) \tag{3.4.1}$$

where τ is a fixed number. If $N_u^*(t)$, $N_u^*(E)$ denote the number of such upcrossings in $(0, t)$, and in the set E respectively, then clearly

$N_u^*(t) = N_u((\tau/\mu)t)$ and $N_u^*(B) = N_u((\tau/\mu)B)$(where $\alpha B = \{\alpha x : x \in B\}$). Further,

$$EN_u^*(1) = EN_u(\tau/\mu) = (\tau/\mu)EN_u(1) = \tau,$$

so that the normalization has provided a constant intensity for our point process. Now writing $T = \tau/\mu$, (3.3.7) may be restated as

$$P\{M(\tau/\mu) \leqslant u\} \to e^{-\tau} \qquad \text{as } u \to \infty. \tag{3.4.2}$$

If $M^*(t)$, $M^*(E)$ denote the quantities related to ξ^* as $M(t)$, $M(E)$ are related to ξ, we see (replacing τ by $h\tau$) that (3.4.2) leads to $P\{M^*(h) \leqslant u\} \to e^{-h\tau}$ and hence by stationarity

$$P\{M^*(I) \leqslant u\} \to e^{-h\tau} \tag{3.4.3}$$

for any interval I of length h. Now since if, e.g., $I = [a, b]$,

$$P\{M^*(I) \leqslant u\} \leqslant P\{N_u^*(I) = 0\} \leqslant P\{M^*(I) \leqslant u\} + P\{\xi^*(a) > u\}$$

and the last term tends to zero as $u \to \infty$ we have from (3.4.3)

$$P\{N_u^*(I) = 0\} \to e^{-h\tau} \qquad \text{as } u \to \infty \tag{3.4.4}$$

for any fixed interval I of length h. That is, we have the limiting Poisson probability for zero events in a fixed interval. The following main theorem shows that this holds not only for the probability of zero events, but for k events, for joint probabilities of $k_1, \ldots, k_n$ events in intervals $I_1, \ldots, I_n$, and indeed in the full weak convergence sense (cf. Theorem 2.4.2).

THEOREM 3.4.1: *Suppose the stationary normal process $\xi(t)$ satisfies* (3.2.1) *and any condition implying* (3.3.5), *such as $r(t) \log t \to 0$ as $t \to \infty$. Then $N_u^* \xrightarrow{d} N$ as $u \to \infty$, where N_u^* is the point process, defined above, consisting of the exceedances of u by $\xi^*(t)$ (i.e., by ξ after the time scale change* (3.4.1)), *and N is a Poisson process with intensity τ. In particular if I is a fixed interval of length h,*

$$P\{N_u^*(I) = k\} \to e^{-\tau h}(\tau h)^k/k! \qquad \textit{as } u \to \infty$$

with corresponding products of Poisson probabilities for the limits of the joint distributions of $N(I_1), \ldots, N(I_r)$ for disjoint intervals $I_1, \ldots, I_r$.

Proof: To prove this convergence, it is sufficient to consider sequences of u values and hence by Theorem 2.4.1 to show that (i) (3.4.4) holds when I is replaced by a finite union $B = \bigcup_1^r I_j$ of r fixed disjoint intervals $I_1, \ldots, I_r$, and (ii) $EN_u^*([a, b]) \to EN([a, b]) = \tau(b - a)$.

(ii) follows at once from the fact that $EN_u^*(1) = \tau$, and requires no further proof.

To prove (i) we note that the same arguments as used in proving Lemmas 3.3.4 and 3.3.5 apply with very little change (and are even simpler here if we take the intervals $I_1, \ldots, I_r$ to be mutually separated by positive distances as we may do) to show that

$$P\left\{\bigcap_{k=1}^{r}\left(M\left(\frac{T}{\mu}I_k\right) \leqslant u\right)\right\} - \prod_{k=1}^{r} P\left\{M\left(\frac{T}{\mu}I_k\right) \leqslant u\right\} \to 0.$$

(Here we have a fixed number of intervals of increasing length whereas there we considered an increasing number of intervals of fixed length, but the mechanics are the same.) This at once translates to give

$$P\{M^*(B) \leqslant u\} - \prod_{k=1}^{r} P\{M^*(I_k) \leqslant u\} \to 0,$$

from which (i) follows simply along the same lines as given above for a single interval I.

As noted above we may obtain the asymptotic distribution of the kth largest local maximum of $\xi(t)$ in $0 \leqslant t \leqslant T$, as a corollary of this result—in a manner analogous to that for the kth largest member of a sequence. We now briefly indicate this procedure.

By a local maximum, we mean a downcrossing of zero by the derivative $\xi'(t)$. $\xi'(t)$ has the covariance function $-r''(\tau)$, and to discuss these crossings we shall need $-r''(\tau)$ to have a finite second derivative. That is, we shall assume that

$$r(\tau) = 1 - \lambda_2\tau^2/2 + \lambda_4\tau^4/4! + o(\tau^4) \qquad \text{as } \tau \to 0. \quad (3.4.5)$$

It is intuitively clear that the local maxima of heights greater than u must (at least for large u) be rather closely related to the upcrossings of u by $\xi(t)$. Certainly at least one local maximum must occur between any two such upcrossings and, in fact, as u increases, these local maxima correspond more and more "in a 1-1 fashion" to the up-crossings, as will shortly be made explicit. Thus we may expect to approximate the probability of at least k local maxima exceeding u by the probability of k upcrossings doing so.

Let $N_u'(T)$ denote the number of local maxima of $\xi(t)$ with heights greater than u, in $0 \leqslant t \leqslant T$ (and $N_u(T)$ the number of u-upcrossings as before). Then under (3.4.5) some calculation along the lines of that in Section 3.2 gives

$$EN_u'(T) = \frac{T}{2\pi}\left(\frac{\lambda_4}{\lambda_2}\right)^{1/2}\left\{1 - \Phi\left(u\left(\frac{\lambda_4}{\Delta}\right)^{1/2}\right) + \lambda_2\left(\frac{2\pi}{\lambda_4}\right)^{1/2}\phi(u)\Phi\left(\frac{\lambda_2 u}{\Delta^{1/2}}\right)\right\}$$

where $\Delta = \lambda_4 - \lambda_2^2$. (A more detailed derivation of this is given in [23].)

It may be shown from this formula that if $T \to \infty$, $\mu T \to \tau$ ($\mu = EN_u(0, 1)$, as always) then $EN_u'(T) \to \tau$ so that $E\{N_u'(T) - N_u(T)\} \to 0$. By using the fact that either $N_u'(T) \geqslant N_u(T)$ or $\xi(T) > u$, some obvious calculation shows that

$$E|N_u'(T) - N_u(T)| \to 0,$$

from which it follows that, for any k,

$$P\{N_u'(T) < k\} - P\{N_u(T) < k\} \to 0, \tag{3.4.6}$$

since the modulus of this difference does not exceed

$$P\{|N_u'(T) - N_u(T)| \geqslant 1\} \leqslant E|N_u'(T) - N_u(T)|.$$

Armed with this we may at once obtain the following results, in which $M_T^{(k)}$ will denote the kth largest local maximum of $\xi(t)$ in $0 \leqslant t \leqslant T$.

THEOREM 3.4.2: *Suppose that the covariance function $r(\tau)$ of $\xi(t)$ satisfies* (3.4.5) *and that $r(t)\log t \to 0$ as $t \to \infty$. Then if $\mu T = \tau$ we have, for any positive integer k,*

$$P\{M_T^{(k)} \leqslant u\} \to e^{-\tau}\sum_{j=0}^{k-1} \tau^j/j! \qquad \textit{as } T \to \infty. \tag{3.4.7}$$

Further

$$P\{a_T(M_T^{(k)} - b_T) \leqslant x\} \to \exp(-e^{-x})\sum_{j=0}^{k-1} e^{-jx}/j! \tag{3.4.8}$$

where a_T and b_T are given by (3.3.9).

Proof: In a similar way to that used for sequences we note that the events $\{M_T^{(k)} \leqslant u\}$ and $\{N'_u(T) \leqslant k - 1\}$ are identical, so that by (3.4.6)

$$P\{M_T^{(k)} \leqslant u\} - P\{N_u(T) \leqslant k - 1\} \to 0.$$

But by Theorem 3.4.1

$$P\{N_u(T) \leqslant k - 1\} = \sum_{j=0}^{k-1} P\{N_u^*(1) = j\} \to e^{-\tau} \sum_{j=0}^{k-1} \tau^j/j!$$

as required for (3.4.7). (3.4.8) then follows at once in the same way as Theorem 3.3.8 followed from Theorem 3.3.7.

3.5. More general assumptions. The basic assumption of the previous section was that the covariance function $r(\tau)$ of our stationary normal process $\xi(t)$ had an expansion

$$r(\tau) = 1 - \lambda_2\tau^2/2 + o(\tau^2) \qquad \text{as } \tau \to 0.$$

As noted in Section 3.2 a more general type of covariance function has the form (3.2.2), i.e.,

$$r(\tau) = 1 - C|\tau|^\alpha + o(|\tau|^\alpha) \qquad (0 < \alpha \leqslant 2). \qquad (3.5.1)$$

For example this includes covariances of the form $e^{-|\tau|^\alpha}$, the case $\alpha = 1$ being that of the Ornstein-Uhlenbeck process.

It is clear that if $\alpha < 2$ we cannot expect a Poisson result to hold for the upcrossings of a high level, since r is not twice differentiable, i.e., $\lambda_2 = \infty$, and the mean number of crossings of any level in unit time is infinite. However, it is certainly possible for the maximum $M(T)$ to have a limiting distribution. The point is that while upcrossings may be numerous (or infinite in number, even), (3.5.1) is sufficient to guarantee continuity of the paths and ensure that $M(T)$ is well defined and finite.

Results of this kind have been obtained by Berman [6], Pickands [30], Qualls and Watanabe [32]. Various methods are used—none being nearly as straightforward as that (as described here) when $\lambda_2 < \infty$.

The result analogous to Theorem 3.3.7 is the following.

THEOREM 3.5.1. *Let $\xi(t)$ be a stationary normal process with zero mean, unit variance, and covariance function $r(t)$ satisfying* (3.5.1) *and* $r(t)\log t \to 0$ *as* $t \to \infty$. *Then for any fixed* τ,

$$P\{M(\tau/\mu) \leqslant u\} \to e^{-\tau}$$

if $\mu \sim C^{1/\alpha} H_\alpha u^{(2/\alpha)-1} \phi(u)$, H_α *being a constant for fixed* α.

As will be seen shortly the quantity μ occurring here is a generalization of the intensity of upcrossings used in Theorem 3.3.7. From this result we obtain the following, just as Theorem 3.3.8 follows from Theorem 3.3.7.

THEOREM 3.5.2: *Let $\xi(t)$ be a stationary normal process with zero mean and unit variance, and let its covariance function $r(\tau)$ satisfy* (3.5.1) and $r(t)\log t \to 0$ *as* $t \to \infty$. *Then*

$$\Pr\{a_T(M(0,T) - b_T) \leqslant x\} \to e^{-e^{-x}} \qquad \text{as } T \to \infty \qquad (3.5.2)$$

where

$$a_T = (2\log T)^{1/2}$$

$$b_T = (2\log T)^{1/2} + (2\log T)^{-1/2}$$

$$\times \left(\left(\frac{1}{\alpha} - \frac{1}{2}\right)\log\log T + \log\left[(2\pi)^{-1/2} C^{-1/\alpha} H_\alpha 2^{(2-\alpha)/2\alpha}\right]\right),$$

H_α *being a certain, strictly positive constant which depends only on* α.

We shall not embark on a proof of Theorem 3.5.1, but just note a few points. It is possible to prove it (cf., e.g., [30], [32]) by using methods which are akin to those for the $\alpha = 2$ case considered above, in their main features, but the details are much more complicated. For example the number of upcrossings of a level u now has infinite expectation when $\alpha < 2$, but we may instead consider the so-called ϵ-upcrossings introduced in [29]. Given $\epsilon > 0$ an ϵ-upcrossing is simply any upcrossing which is preceded by an interval of length at least ϵ, containing no other upcrossings. Clearly the number $N_{\epsilon,u}(t)$ of ϵ-upcrossings in $(0, t)$ is finite (indeed bounded by t/ϵ) and has moments of all orders.

A basic step in most proofs of Theorem 3.5.1 is again to obtain an expression for the tail of the distribution of the maximum in a *fixed* interval corresponding to Lemma 3.3.3, where $\alpha = 2$. Specifically the following result holds.

LEMMA 3.5.2: *Suppose that* (3.5.1) *holds. For any fixed* $\epsilon > 0$ *write* $\mu_\epsilon = EN_{\epsilon,u}(1)$. *Then for any* $h > 0$

$$P\{M(h) > u\} \sim \mu_\epsilon h \qquad \text{as } u \to \infty \tag{3.5.3}$$

where also

$$\mu_\epsilon \sim C^{1/\alpha} H_\alpha u^{(2/\alpha)-1} \phi(u). \tag{3.5.4}$$

Even though the details are different, one proof of this result makes use of a comparison, not with the harmonic process used when $\alpha = 2$, but with a certain nonstationary normal process. Proofs of the main result may then be based on (3.5.3) in a manner reminiscent of (but more complicated than) the derivation for the case $\alpha = 2$ (cf. [29] or [6] for details). Further, since a number of the arguments used in the case $\alpha = 2$ do generalize to this case, it seems likely that it might well be possible to obtain a simpler and more unified proof to include the case $\alpha < 2$, along the lines of that given above for $\alpha = 2$.

Having obtained the basic results for the maximum, we may use Theorem 2.4.1 again as we did in Section 3.4, and obtain a Poisson limit for the point process of ϵ-upcrossings, with corresponding corollaries as before.

Further, a little more generality may be obtained by including a function of "slow growth" as the coefficient of $|\tau|^\alpha$ in (3.5.1), and some of the results extend to apply to processes with stationary increments. For such extensions and associated questions we refer, e.g., to [32], [7].

Finally we note again that our object in this paper is to show how parts of extreme value theory may be extended from the classical i.i.d. case, to include cases including not only strong mixing but also much weaker forms of dependence. We have been very selective in our choice of topics for this purpose and many interesting areas (e.g., stability, laws of large numbers, extremal processes, excursions above levels, and so on) have not been mentioned at all. This does not imply that these topics are in any way less interesting—simply that they lie outside the scope and purposes of the present paper.

Acknowledgments. It is a pleasure to thank Dr. H. Rootzen, Dr. G. Lindgren, Professor M. Rosenblatt, and Dr. V. Watts for their valuable comments and suggestions concerning this and related work. I wish also to record my appreciation to the Office of Naval Research, which sponsored (under Contract N00014-75C-0809) some of the research recorded here.

Note added in proof. Since the writing of this paper, the converse question raised in several places concerning Theorem 2.3.1 has been settled by Mr. R. Davis, who has shown that (2.3.2) also implies (2.3.1) in that Theorem. I am very grateful to Mr. Davis for an advance copy of his paper [48] which will appear in the Annals of Probability. (See also [23] for a somewhat different version of this result.)

REFERENCES

(This list contains some related works not referred to in the text.)

1. R. R. Bahadur, "A note on quantiles in large samples," *Ann. Math. Statist.*, **37** (1966), 577–580.
2. Ban Cheng, "The limiting distributions of order statistics," *Chinese Math.*, **6** (1965), 84–104.
3. Yu. K. Belayev, "On the number of intersections of a level by a Gaussian stochastic process II," *Teor. Ver. i Prim.*, **12** (1967), 444–457.
4. S. M. Berman, "Limit theorems for the maximum term in stationary sequences," *Ann. Math. Statist.*, **35** (1964), 502–516.
5. ———, "Asymptotic independence of the numbers of high and low level crossings of stationary Gaussian processes," *Ann. Math. Statist.*, **42** (1971), 927–945.
6. ———, "Maxima and high level excursions of stationary Gaussian processes," *Trans. Amer. Math. Soc.*, **160** (1971), 67–85.
7. ———, "Maximum and high level excursions of a Gaussian process with stationary increments," *Ann. Math. Statist.*, **43** (1972), 1247–1266.
8. D. Brillinger, "On the number of solutions of systems of random equations," *Ann. Math. Statist.*, **43** (1972), 534–540.
9. D. M. Chibisov, "On limit distributions of order statistics," *Theor. Probability Appl.*, **9** (1964), 142–148.

10. H. Cramér and M. R. Leadbetter, *Stationary and Related Stochastic Processes*, Wiley, New York, 1967.

11. C. M. Deo, "A note on strong mixing Gaussian sequences," *Ann. Prob.*, **1** (1973), 186–187.

12. A. Dvoretzky, "Asymptotic normality for sums of dependent random variables," *Proc. 6 Berkeley Symp. on Stat. and Prob.*, **2** (1972), 513–535.

13. R. A. Fisher and L. H. C. Tippett, "Limiting forms of the frequency distribution of the largest or smallest member of a sample," *Proc. Cambridge Philos. Soc.*, **24** (1928), 180–190.

14. J. Galambos, "On the distribution of the maximum of random variables," *Ann. Math. Statist.*, **43** (1972), 516–521.

15. ———, "A general Poisson limit theorem of probability theory," *Duke Math. J.*, **40** (1973), 581–586.

16. B. V. Gnedenko, "Sur la distribution limite du terme maximum d'une série aléatoire," *Ann. Math.*, **44** (1943), 423–453.

17. B. V. Gnedenko and A. N. Kolmogorov, *Limit Distributions for Sums of Independent Random Variables*, Addison-Wesley, New York, 1954.

18. E. J. Gumbel, *Statistics of Extremes*, Columbia University Press, New York, 1958.

19. L. de Haan, "On regular variation and its application to the weak convergence of sample extremes," *Amsterdam Math. Centre Tract 32*, (1970).

20. O. Kallenberg, *Random Measures*, Akad. Ver. (D.D.R.) Berlin, 1975.

21. M. R. Leadbetter, "On extreme values in stationary sequences," *Z. Wahr. verw. Geb.*, **28** (1974), 289–303.

22. ———, "Weak convergence of high level exceedances by a stationary sequence," *Z. Wahr. verw. Geb.*, **34** (1976), 11–15.

23. M. R. Leadbetter, G. Lindgren, and H. Rootzen, "Extremal and related properties of stationary processes," University of Umeå Technical Report, to appear (revised form of lecture notes, University of Lund, 1974), Umeå, Sweden, 1978.

24. G. Lindgren, J. De Maré, and H. Rootzen, "Weak convergence of high level crossings and maxima for one or more Gaussian processes," *Ann. Prob.*, **3** (1975), 961–978.

25. R. M. Loynes, "Extreme values in uniformly mixing stationary stochastic processes," *Ann. Math. Statist.*, **36** (1965), 993–999.

26. T. Mori, "Limit laws for maxima and second maxima from strong mixing processes," *Ann. Prob.*, **4** (1976), 122–126.

27. G. L. O'Brien, "Limit theorems for the maximum term of a stationary process," *Ann. Prob.*, **2** (1974), 540–545.

28. J. Tiago de Oliveira, "Extremal processes: definition and properties," *Publ. Inst. Statist. Univ. Paris*, **17** (1968), 25–36.

29. J. Pickands, "Upcrossing probabilities for stationary Gaussian processes," *Trans. Amer. Math. Soc.*, **145** (1969), 51–73.

30. ———, "Asymptotic properties of the maximum in a stationary Gaussian process," *Trans. Amer. Math. Soc.*, **145** (1969), 75–86.

31. ———, "An iterated logarithm law for the maximum in a stationary Gaussian sequence," *Z. Wahr. verw. Geb.*, **12** (1969), 344–353.

32. C. Qualls and H. Watanabe, "Asymptotic properties of Gaussian processes," *Ann. Math. Statist.*, **43** (1972), 580–596.

33. S. I. Resnick and M. Rubinovitch, "The structure of extremal processes," *Adv. Appl. Prob.*, **5** (1973), 287–307.

34. M. Rosenblatt, "Dependence and asymptotic independence for random processes," this volume.

35. D. Slepian, "The one-sided barrier problem for Gaussian noise," *Bell Syst. Tech. J.*, **41** (1962), 463–501.

36. N. V. Smirnov, "The limit distributions for the terms of a variational series," *Amer. Math. Soc. Transl.*, **67** (1952).

37. ———, "Some remarks on limit laws for order statistics," *T. Prob. Appl.*, **12** (1967), 336–337.

38. V. A. Volkonski and Y. A. Rozanov, "Some limit theorems for random functions II," *Teor. Ver. Prim.*, **6** (1961), 202–215.

39. G. S. Watson, "Extreme values in samples from m-dependent stationary stochastic processes," *Ann. Math. Statist.*, **25** (1954), 798–800.

40. V. Watts, "Limit theorems and representations for increasing rank order statistics from dependent sequences," Ph.D. thesis, Dept. of Statistics, University of North Carolina at Chapel Hill, 1977.

41. R. E. Welsch, "A weak convergence theorem for order statistics from strong-mixing processes," *Ann. Math. Statist.*, **42** (1971), 1637–1646.

42. ———, "Limit laws for extreme order statistics from strong mixing processes," *Ann. Math. Statist.*, **43** (1972), 439–446.

43. Wu Chuan-Yi, "The types of limit distributions for some terms of variational series," *Sci. Sinica*, **15** (1966), 749–762.

Sample references concerning extensions and applications:

44. R. J. Adler, "Weak convergence results for extremal processes generated by dependent random variables," to appear in *Ann. Prob.*

45. R. J. Adler and A. M. Hasofer, "Level crossings for random fields," *Ann. Prob.*, **4** (1976), 1–12.

46. P. J. Bickel and M. Rosenblatt, "Two-dimensional random fields," *Multivariate Analysis III*, Krishnaiah ed., (1973), 3–15.
47. ———, "On some global measures of the deviations of density function estimates," *Ann. Statist.*, **1** (1973), 1071–1095.
48. R. Davis, "Maxima and minima of stationary sequences," to appear in *Ann. Prob.*
49. D. G. Kendall, "Hunting quanta," *Philos. Trans. Roy. Soc. London Ser. A*, **276** (1974), 231–266.

ASYMPTOTIC ANALYSIS OF STOCHASTIC EQUATIONS*

G. C. Papanicolaou

1. INTRODUCTION

The purpose of this survey is to collect together and present with a minimum of mathematical detail some problems and results on asymptotics for stochastic equations. Since the subject is very large and rapidly expanding, the examples presented here deal only with one general problem: the effect of rapid, noisy fluctuations on systems (ordinary differential equations).

More specifically, we consider questions (in the context of examples) like the following. How does one measure the effect of noise on systems (scaling)? How do we identify the situations (asymptotic limits) where noise phenomena are fully developed (white noise)?

* Research supported by the National Science Foundation under Grant No. MCS75-09837.

How do we develop a "fluctuation theory" for solutions of differential equations with random parameters? How do we treat problems where noise phenomena are fully developed but have a weak effect?

In Sections 4 to 16 we give some answers to these and a few other related questions. We also provide references to the extensive literature.

2. BRIEF SURVEY OF SOME FACTS FROM PROBABILITY

We refer to [1]–[5] for systematic exposition of the theory of stochastic processes.

A real valued stochastic process with parameter set $T = [0, \infty)$ or $T = \{1, 2, 3, \ldots\}$ is a family $\{x(t),\ t \in T\}$ of real valued random variables, i.e., a consistent family of distributions $\{F_{t_1,t_2,\ldots,t_n}(x_1, x_2, \ldots, x_n)\}$ where for each $t_1, t_2, \ldots, t_n \in T$ the joint distribution function of $x(t_1), x(t_2), \ldots, x(t_n)$ is given by the corresponding member of this family of distributions. Consistent means that marginal distributions match when the marginal index set $\{t_1, t_2, \ldots, t_n\}$ is the same.

Given such a stochastic process one can then construct a probability space $(\Omega, \mathcal{F}, P)$ (where Ω is a set, $\mathcal{F}$ is a σ-algebra of subsets, and P is a probability measure on $\mathcal{F}$) and functions $x(t, \omega)$ on $T \times \Omega$ into R^1 such that

$$P\{\omega \in \Omega : x(t_1) \in A_1, x(t_2) \in A_2, \ldots, x(t_n) \in A_n\}$$
$$= \int_{A_1} \cdots \int_{A_n} dF_{t_1,t_2,\ldots,t_n}(x_1, x_2, \ldots, x_n).$$

Here, $t_1, t_2, \ldots, t_n \in T$ and $A_1, A_2, \ldots, A_n$ are Borel subsets of R^1. The space Ω can be taken to be the space of all functions on T in which case for $\omega \in \Omega$, $x(t, \omega)$ is the value at t of the function ω.

If the process $\{x(t), t \in T\}$ is stochastically continuous, i.e.,

$$\lim_{h \downarrow 0} P\{|x(t + h) - x(t)| > \delta\} = 0, \qquad \text{for all } \delta > 0,$$

then the process can be chosen so that $x(t, \omega)$ is a jointly measurable function of t and ω and the process is separable. This means that the probability of events that involve uncountably many t's is the same when only a countable dense set of t's is used.

Let $\mathscr{F}_t$ be the σ-algebra generated by events that involve the path $x(s, \omega)$ only for times $0 \leqslant s \leqslant t$. Clearly the σ-algebras $\mathscr{F}_t$ increase with t and they are contained in $\mathscr{F}$.

Given a random variable Y on Ω, i.e., a real valued measurable function on Ω such that

$$E\{|Y|\} = \int_\Omega |Y(\omega)| P(d\omega) < \infty,$$

we define the conditional expectation $E\{Y|\mathscr{F}_t\}$ of Y given $\mathscr{F}_t$ as that $\mathscr{F}_t$ measurable random variable for which

$$E\{\chi_B Y\} = E\{\chi_B E\{Y|\mathscr{F}_t\}\}$$

for any event $B \in \mathscr{F}_t$. Here $\chi_B = \chi_B(\omega)$ is the characteristic function of the set B (equal to one if $\omega \in B$ and zero if $\omega \notin B$). Conditional expectations are always well defined up to P almost everywhere equivalence. Note that

$$E\{Y\} = E\{E\{Y|\mathscr{F}_t\}\}.$$

The basic classes of processes are the Markov processes, the martingales, the Gaussian processes and the stationary processes, among others. A process $\{x(t), t \geqslant 0\}$ is called Markov if the conditional probability of an event that involves the paths for times $s \geqslant t + h$ given the past up to time t, is a function of the present state $x(t)$. More formally, if $B \in \mathscr{F}$ and $B \notin \mathscr{F}_{t+h}$, then $E\{\chi_B|\mathscr{F}_t\}$, which is in general a functional of $x(s)$ for $0 \leqslant s \leqslant t$, is a point function of $x(t)$ for Markov processes. This means that the measure P on $(\Omega, \mathscr{F})$ (which is what is meant by the process) has relatively simple structure and is in fact determined by the initial distribution

$$P\{x(0) \leqslant x\} = F_0(x)$$

and the transition probability distributions

$$P\{x(t) \leqslant y \mid x(s) = x\} = F(s, t, x, y).$$

Martingales are processes (measures) for which

$$\sup_{t \geq 0} E\{|x(t)|\} < \infty$$

and

$$E\{x(t)|\mathscr{F}_s\} = x(s), \qquad 0 \leqslant s \leqslant t$$

almost everywhere with respect to P. The importance of this class of processes will become apparent shortly.

Gaussian processes are characterized by the fact that the finite dimensional distributions $F_{t_1,\ldots,t_n}(x_1,\ldots,x_n)$ are all Gaussian. Hence the measure P is completely characterized by the mean function

$$m(t) = E\{x(t)\}$$

and the covariance function

$$\rho(t, s) = E\{(x(t) - m(t))(x(s) - m(s))\}, \qquad t, s \geqslant 0.$$

Stationary processes are characterized by the property that for any $t_1, t_2, \ldots, t_n \in T$ and $h > 0$, $x(t_1), \ldots, x(t_n)$, and $x(t_1 + h), \ldots, x(t_n + h)$ have the same distribution. A Gaussian process is stationary if and only if the mean function is independent of t and the covariance a function of $|t - s|$ only.

A positive random variable τ on Ω, the sample space of a process, is called a stopping time if the events $\{\tau \leqslant t\}$ are $\mathscr{F}_t$ measurable. If $\tau \wedge t = \min(t, \tau)$, the process $x(\tau \wedge t)$ is called the stopped process corresponding to $x(t)$. If τ is a stopping time, the σ-algebra of events $A \cap \{\tau \leqslant t\}$ with $A \in \mathscr{F}$ and $t \geqslant 0$ is denoted by $\mathscr{F}_\tau$. It represents events associated with the trajectories up to the random time τ. If $\sigma \leqslant \tau$ are two bounded stopping times and $x(t)$ is a martingale then

$$E\{x(\tau)|\mathscr{F}_\sigma\} = x(\sigma) \qquad \text{a.s. } P,$$

i.e., the martingale property holds also for random times. This is the optional stopping theorem.

Let us now consider Markov processes in more detail. Clearly it is not necessary to restrict oneself to real valued processes, so we shall assume that the state space of $\{x(t), t \geqslant 0\}$ is some set S which is usually a complete separable metric space. A time homogeneous Markov process on S is characterized by the following: (i) the initial distribution $\mu(A)$ which is a probability measure on the Borel subsets Σ of S (i.e., $A \in \Sigma$) and (ii) the transition probability function $P(t, x, A)$, $t \geqslant 0$, $x \in S$, $A \in \Sigma$, which is a measurable function of t

and x for each A and a probability measure for each t and x. Moreover, the Chapman-Kolmogorov equation holds

$$P(t+s, x, A) = \int_S P(t, x, dy)P(s, y, A).$$

The process $\{x(t), t \geqslant 0\}$ is now constructed so that

$$P\{x(t_1) \in A_1, x(t_2) \in A_2, \ldots, x(t_n) \in A_n\}$$
$$= \int \mu(dx) \int_{A_1} P(t_1, x, dy_1) \int_{A_2} P(t_2 - t_1, y_1, dy_2) \cdots$$
$$\times \int_{A_n} P(t_n - t_{n-1}, y_{n-1}, dy_n).$$

Let $B(S, \Sigma)$ be the Banach space of bounded measurable functions on S with the sup norm and let $M(S, \Sigma)$ be the Banach space of finite signed measures with the total variation norm. The transition function induces a semigroup of operators $T(t)$ on B and M as follows:

$$T(t)f(x) = \int_S P(t, x, dy)f(y), \qquad f \in B,$$

$$\mu T(t)(A) = \int_S \mu(dx)P(t, x, A), \qquad \mu \in M.$$

The semigroup $T(t)$ is called the transition semigroup. It has the additional property that it is positivity preserving: $T(t)f \geqslant 0$ if $f \geqslant 0$, $\mu T(t) \geqslant 0$ if $\mu \geqslant 0$, and $T(t)1 = 1$ where 1 is the function identically equal to 1 on S. The set of functions for which

$$\lim_{t \downarrow 0} T(t)f = f$$

is called the strongly continuous center of $T(t)$. The set of functions for which

$$\lim_{t \downarrow 0} \frac{1}{t}(T(t)f - f) = g$$

exists is called the domain of the infinitesimal generator A of $T(t)$ and we set $Af = g$. The domain is denoted by $D(A)$.

Let S be a compact set, for simplicity. Let $C(S)$ be the Banach space of bounded continuous functions on S. If $T(t): C(S) \to C(S)$, it is called a Feller semigroup and the corresponding process a Feller process. If $T(t): B(S) \to C(S)$, then it is called a strong Feller semigroup. If

$$\sup_x P(t, x, \{d(x, y) > \delta\}) = 0(t), \qquad t \downarrow 0, \qquad \delta > 0,$$

where $d(\cdot, \cdot)$ is the distance function, then the trajectories of the process do not have, with probability one, discontinuities of the second kind. They can thus be chosen to be right continuous (with left-hand limits). A right continuous Feller process is also a strong Markov process, i.e., the Markov property is valid not only when conditioning with respect to a fixed time t but also when conditioning with respect to a stopping time τ.

The domain $D(A)$ of the generator of $T(t)$ is dense in the strongly continuous center. Moreover if

$$R_\lambda = \int_0^\infty e^{-\lambda t}\, T(t)\, dt, \qquad \lambda > 0$$

is the resolvent operator then, $R_\lambda : D(A) \to$ the strongly continuous center and $\|\lambda R_\lambda\| \leqslant 1$. Conversely (Hille-Yoshida theorem) if A is densely defined in a Banach space, $(\lambda - A): D(A) \to$ the Banach space for $\lambda > 0$ and $\lambda\|g\| \leqslant \|(\lambda - A)g\|$ for $g \in D(A)$, then A is the infinitesimal generator of a strongly continuous semigroup on the Banach space (all semigroups are contractions $\|T(t)\| \leqslant 1$).

Since A, the infinitesimal generator, is what is usually given (the coefficients in the Fokker-Plank equation, for example, or the birth and death rates), a theorem such as the above describes when the "coefficients" determine the process uniquely.

Suppose $T(t)$ is strongly continuous on $C(S)$, S is compact, and let $x(t)$, the corresponding process, be right continuous. Suppose $f(x)$ is in $D(A)$ and $Af(x) = 0$ (f is then called harmonic). Then the real valued process $f(x(t))$ is a martingale because

$$\begin{aligned} E\{f(x(t+s))|\mathscr{F}_s\} &= T(t)f(x(s)) \qquad \text{a.s.} \\ &= f(x(s)) + \int_0^t T(\sigma)Af(x(s))\, d\sigma \\ &= f(x(s)). \end{aligned}$$

More generally, if $f(\cdot) \in D(A)$ then

$$f(x(t)) - \int_0^t Af(x(s))\, ds$$

is a martingale. Similarly, if for each t, $f(t, \cdot) \in D(A)$, $f(t, x)$ is bounded and has bounded t derivative then,

$$f(t, x(t)) - \int_0^t \left(\frac{\partial}{\partial s} + A\right) f(s, x(s))\, ds$$

is a martingale. Also, if $u(x)$ is strictly positive and $u(\cdot) \in D(A)$,

$$u(x(t)) \exp\left\{-\int_0^t \frac{Au(x(s))}{u(x(s))}\, ds\right\}$$

is a martingale.

Let us discuss this last example in more detail. Let $V(x)$ be a continuous function on S and let V denote the operator corresponding to multiplication by $V(x)$. If A is the infinitesimal generator of a semigroup $T(t)$ as above, then $A + V$ is also the generator of a semigroup which we denote by $T_V(t)$. Then we have that

$$T_V(t)f(x) = E_x\left\{\exp\left[\int_0^t V(x(s))\, ds\right] f(x(t))\right\}$$

where E_x denotes expectation relative to the measure P_x in the sample space when the initial measure (at time $t = 0$) is concentrated at the point $x \in S$. To verify the above formula, we assume $f \in D(A)$ and, by direct computation we find that

$$\lim_{t \downarrow 0} \frac{1}{t}(T_V(t)f - f) = Af + Vf,$$

where we use the right-continuity of the paths. Since $A + V$ generates (uniquely) a semigroup and $T_V(t)$ as defined above is a semigroup, the identification is complete.

Suppose now that $u(x)$ is strictly positive and $u(\cdot) \in D(A)$. Then $V(x) = -Au(x)/u(x)$ is continuous and so

$$\begin{aligned} T_V(t)u(x) &= E_x\left\{\exp\left[-\int_0^t \frac{Au(x(s))}{u(x(s))}\, ds\right] u(x(t))\right\} \\ &\equiv u(x) \end{aligned}$$

because $dT_V(t)u/dt \equiv 0$ as is easily shown. From this the martingale property follows immediately.

The above examples show that martingales are produced in profusion in the analysis of Markov processes (and in many other situations).

Among the basic properties of martingales, in addition to the optional stopping theorem, we mention the martingale convergence theorems [2] and various important inequalities for them, the simplest being Kolmogorov's inequality ($x^+ = x \vee 0 = \max(x, 0)$)

$$P\left\{\sup_{0 \leqslant t \leqslant T} x(t) > \delta\right\} \leqslant \frac{E\{x^+(T)\}}{\delta}, \qquad \delta > 0.$$

The systematic use of martingales to study Markov processes (principally diffusion processes) is carried out in [6], [7].

Diffusion processes on R^n, say, are Markov processes with continuous trajectories such that if $P(t, x, A)$ is their transition probability function, it satisfies

$$\text{(i)} \lim_{t \downarrow 0} \sup_x \frac{1}{t} \int_{|x-y| > \delta} P(t, x, dy) = 0,$$

$$\text{(ii)} \lim_{t \downarrow 0} \sup_x \left[\frac{1}{t} \int_{|x-y| \leqslant \delta} (x - y)P(t, x, dy) - b(x)\right] = 0,$$

$$\text{(iii)} \lim_{t \downarrow 0} \sup_x \left[\frac{1}{t} \int_{|x-y| \leqslant \delta} (x - y) \otimes (x - y)P(t, x, dy) - a(x)\right] = 0.$$

In (i)–(iii) $\delta > 0$ is arbitrary and $b(x)$ and $a(x)$ are n-vector and $n \times n$ matrix functions respectively. Condition (i) is called Lindeberg's condition and it implies continuity of trajectories along with the Feller property for $P(t, x, dy)$. From (i)–(iii) one finds easily that if $f(x)$ is smooth and bounded, then

$$T(t)f(x) = E_x\{f(x(t))\} = \int P(t, x, dy)f(y)$$

where

$$A = \frac{1}{2} \sum_{i,j=1}^{n} a_{ij}(x) \frac{\partial^2}{\partial x_i \partial x_j} + \sum_{j=1}^{n} b_j(x) \frac{\partial}{\partial x_j}.$$

Given infinitesimal diffusion coefficients $a(x) = (a_{ij}(x))$ and infinitesimal drift coefficients $b(x) = (b_j(x))$, when does the operator A above determine uniquely the semigroup $T(t)$ and hence the process ($\equiv$measure on the space of continuous trajectories in R^n)? This can be answered by applying the Hille-Yoshida theorem, the difficult thing being the description of the range of the resolvent $(\lambda - A)^{-1}$, $\lambda > 0$. Such questions can be analyzed by a variety of methods, more or less probabilistic. If the coefficients $(a_{ij}(x))$ are bounded continuous and uniformly positive definite

$$\sum_{i,j=1}^{n} a_{ij}(x)\xi_i\xi_j \geqslant \alpha \sum_{i=1}^{n} \xi_i^2, \qquad \alpha > 0,$$

and if the $b_j(x)$ are bounded and measurable one can construct a unique probability measure P_x on $C([0, \infty), R^n)$ for each $x \in R^n$ such that

(i) $P_x\{x(0)) = x\} = 1$,

(ii) $f(x(t)) - \int_0^t Af(x(s))\,ds$ is a P_x martingale,

for each $f(x)$ smooth and of compact support [6]. This is the martingale approach to diffusions and it provides, of course, a semigroup $T(t)$ which is Feller. If the coefficients are smooth and bounded, ellipticity, not uniform ellipticity, is enough.

Let $\mathscr{D} \subset R^n$ be a bounded open set and let τ be the first time the diffusion process $x(t)$ with generator A reaches $\partial\mathscr{D}$, the boundary of $\mathscr{D}$. Suppose that the boundary value problem

$$\frac{\partial u(t, x)}{\partial t} = \frac{1}{2}\sum_{i,j=1}^{n} a_{ij}(x)\frac{\partial^2 u(t, x)}{\partial x_i \partial x_j} + \sum_{j=1}^{n} b_j(x)\frac{\partial u(t, x)}{\partial x_j}, \qquad t > 0, \quad x \in \mathscr{D},$$

$$u(0, x) = f(x), \qquad x \in \mathscr{D},$$

$$u(t, x) = g(x), \qquad x \in \partial\mathscr{D}, \qquad t > 0$$

has a smooth bounded solution when the data f and g are smooth.

On applying the optional stopping theorem to the martingale $u(t - \tau \wedge s, x(\tau \wedge s))$, $0 \leqslant s \leqslant t$, ($t$ is fixed) we find that

$$E_x\{u(t - \tau \wedge t, x(\tau \wedge t))\} = u(t, x).$$

Thus,

$$u(t, x) = E_x\{f(x(t))\chi_{\{t<\tau\}}\} + E_x\{g(x(\tau))\chi_{\{\tau \leqslant t\}}\},$$

which shows how the solution of the boundary value problem is expressed probabilistically. Many similar problems can be given a probabilistic representation, including the case of Neumann boundary conditions. For a martingale approach to such problems we refer to [8].

Diffusion processes are of sufficient importance that it is desirable to have many different ways of constructing them from their local parameters $a(x)$, $b(x)$ (and boundary conditions if any). The continuity of the process and the definition of $a(x)$ and $b(x)$ above prompt one to attempt to construct the process as a functional of another, canonical process. The canonical process is Brownian motion where $a(x) = (\delta_{ij})$ and $b(x) \equiv 0$. Thus, for Brownian motion

$$P(t, x, dy) = \frac{e^{-|x-y|^2/2t}}{(2\pi t)^{n/2}}\, dy$$

where $|x - y|^2 = \sum_{i=1}^n (x_i - y_i)^2$. Now the defining relations for $a(x)$ and $b(x)$ above can be written roughly as

$$x(t + \Delta t) - x(t) \sim b(x(t))\Delta t + \sigma(x(t))(w(t + \Delta t) - w(t))$$

where $w(t)$ is the Brownian motion process on R^n and $\sigma(x)$ is the symmetric square root of the infinitesimal diffusion matrix $a(x)$. Clearly the above relation should tend to become an exact equation in the limit $\Delta t \to 0$. We write

$$dx(t) = b(x(t))\, dt + \sigma(x(t))\, dw(t)$$

or in integral form

$$x(t) = x + \int_0^t b(x(s))\, ds + \int_0^t \sigma(x(s))\, dw(s).$$

If one can solve the above equation and express $x(\cdot)$ as a functional of $w(\cdot)$, then the induced transformation of the measure of the Brownian motion process yields the measure of the diffusion process that we want. But we also have more, namely, a pathwise representation of the process as a functional of Brownian motion.

Before attempting to solve the above integral equation, the Brownian stochastic integral must be defined appropriately. The support of the measure of the Brownian motion is contained in the collection of continuous trajectories, but almost all paths are not of bounded variation. Since local properties of w will be reflected in local properties of x the integral

$$\int_0^t \sigma(x(s))\, dw(s)$$

requires a special definition and theory, due to Itô [3, 4].

Briefly, one defines the integral for bounded nonanticipating functionals $f(t, \omega)$ ($\omega \in \Omega$ is a continuous path), i.e., $f(t, \omega)$ is bounded and measurable jointly in (t, ω) and $f(t, \cdot)$ is $\mathscr{F}_t$ measurable for each t, by first defining it for simple nonanticipating functionals

$$\int_0^t f(s, \omega)\, dw(s) = \sum f(t_k, \omega)(w(t_{k+1}) - w(t_k))$$

where the sum is over the finite number of values of the simple functional. Note that the functional is evaluated at the leftmost point of (t_k, t_{k+1}). This makes the stochastic integral a zero-mean martingale and

$$E\left\{\left(\int_0^t f(s, \omega)\, dw(s)\right)^2\right\} = E\left\{\int_0^t f^2(s, \omega)\, ds\right\}.$$

Passage to the limit gives the stochastic integral in general with the martingale property preserved as well as the above identity (f is assumed bounded).

With the stochastic integral and its properties established we return to the stochastic integral equation and attempt to solve it by iteration. As in the case of ordinary differential equations, we must have a Lipschitz condition on $b(x)$ and $\sigma(x)$. If we do, the iteration process converges and the representation of the diffusion process as a nonlinear functional of Brownian motion is accomplished.

Not every diffusion can be so represented however, the difficulty being the absence of the Lipschitz continuity for b or σ. If one wants information contained in the measure only, one does not need the stochastic differential equations. On the other hand, in many problems in modelling and control theory the pathwise representation is important and leads to a number of significant conclusions.

We conclude this brief discussion with some remarks regarding approximations of stochastic processes which is our main topic in Sections 4–16.

One class of approximations concerns behavior of appropriately normalized functionals of a process over $[0, t]$ and then passage to the limit as $t \uparrow \infty$, the convergence being anything from convergence almost everywhere to convergence in the mean to convergence in distribution (weak convergence). For example, if $\{x(t), t \geqslant 0\}$ is a Markov process on a state space S and if $f(x)$ is a bounded measurable function on S, what is the behavior of $\int_0^t f(x(s))\, ds$ for $t \uparrow \infty$? If $x(t)$ is ergodic and $\bar{P}(dy)$ is its stationary distribution (see next section for more detailed discussion), then

$$\frac{1}{t}\int_0^t f(x(s))\, ds \to \int f(x)\bar{P}(dy) \equiv \bar{f}, \qquad t \uparrow \infty$$

for almost all starting points relative to $\bar{P}$. We may then ask about the behavior of

$$\frac{1}{\sqrt{t}}\int_0^t (f(x(s)) - \bar{f})\, ds.$$

Since the limit here (in the simplest case) is a Gaussian random variable, the notion of convergence is weak convergence. However, it is usually of interest, and not much more difficult, to ask for a bit more, as follows.

Let t go to infinity as $\epsilon \to 0$ by letting $t = \tau/\epsilon^2$ with τ an extra parameter. The quantity of interest is then, up to a factor $1/\sqrt{\tau}$,

$$\epsilon\int_0^{\tau/\epsilon^2} (f(x(s)) - \bar{f})\, ds = \frac{1}{\epsilon}\int_0^{\tau} (f(x^{\epsilon}(s)) - \bar{f})\, ds,$$

where $x^{\epsilon}(t) \equiv x(t/\epsilon^2)$. Now we treat this quantity as a family of processes with parameter $\tau \geqslant 0$ and $\epsilon > 0$ labelling the family. That is, we have a family of measures P^{ϵ} on $C([0, \infty), R^1)$. The question above generalizes to showing that P^{ϵ} converges weakly to Brownian motion. By weak convergence of measures we mean the following. If X is a separable metric space and P^{ϵ} is a family of Borel probability measures on X, P^{ϵ} converges weakly to P if

$$\int_X f(x)P^{\epsilon}(dx) \to \int_X f(x)P(dx)$$

for each bounded continuous function f on X. We shall be concerned with weak convergence in most of what follows.

In the context of Markov processes, if the measures P^{ϵ} can be shown to be weakly compact (there are simple sufficient conditions for this [9]) then it suffices to show that the semigroups $T^{\epsilon}(t)$ converge to a limit semigroup in the strong topology as $\epsilon \to 0$, uniformly on compact time intervals. Let A^{ϵ} be the corresponding generator and R_{λ}^{ϵ} the resolvents.

By a well-known theorem [10], if for $\lambda > 0$, $R_{\lambda}^{\epsilon} f$ converges to $R_{\lambda} f$ for $f \in C(S)$, S compact, say, as $\epsilon \to 0$, then the corresponding semigroups converge. Here it is not actually necessary to assume that R_{λ} is the resolvent of a generator; this follows. On the other hand, one usually has access concretely to A^{ϵ} and not R_{λ}^{ϵ} (or $T^{\epsilon}(t)$). If there is a dense set $D \subset \bigcap_{\epsilon > 0} D(A^{\epsilon})$ in the Banach space and a generator A of a semigroup (all semigroups are contractions here) $T(t)$ such that the closure of $(\lambda - A)D$ is the whole space and $A^{\epsilon} f \to Af$ for $f \in D$, then the resolvents converge. For given $\tilde{f}$ in the Banach space there is an f of the form $(\lambda - A)g$, $g \in D$, arbitrarily close to it and, for $\lambda > 0$,

$$\begin{aligned} \|R_{\lambda}^{\epsilon}(\lambda - A)g - R_{\lambda}(\lambda - A)g\| &= \|R_{\lambda}^{\epsilon}(\lambda - A)g - g\| \\ &= \|R_{\lambda}^{\epsilon}(A^{\epsilon} - A)g\| \leqslant \frac{1}{\lambda}\|A^{\epsilon}g - Ag\| \to 0 \end{aligned}$$

as $\epsilon \downarrow 0$.

It is actually enough that there exist a set D as above such that for $f \in D$ there is a sequence $f^{\epsilon} \in D$ and $f^{\epsilon} \to f$, $A^{\epsilon} f^{\epsilon} \to Af$ [11]. To see this we note that with g as above and $g^\epsilon \to g$,

$$\|R_\lambda^\epsilon(\lambda - A)g - R_\lambda(\lambda - A)g\| = \|R_\lambda^\epsilon(A^\epsilon g^\epsilon - Ag) - R_\lambda^\epsilon A^\epsilon(g^\epsilon - g)\|$$

$$\leqslant \frac{1}{\lambda}\|A^\epsilon g^\epsilon - Ag\| + 2\|g^\epsilon - g\| \to 0$$

as $\epsilon \downarrow 0$.

In many situations (most of what follows in Sections 4–16) the measures P^ϵ (or the processes $x^\epsilon(t)$) are not Markovian. However, it may be possible to identify $x^\epsilon(t)$ as the component of a process on a larger state space, this bigger one being Markov. In addition, the limit measure of the P^ϵ turns out to be itself Markovian. In this case one first shows that the P^ϵ are a relatively weakly compact family of measures and then one shows, as above, that the semigroup on the bigger space converges as $\epsilon \to 0$ to a semigroup on the subspace when acting on elements of the subspace. In this contraction of space context, the device of choosing $f^\epsilon \to f$ so that $A^\epsilon f^\epsilon \to Af$ is very useful and will be illustrated by many examples in the following.

3. BRIEF REVIEW OF SOME FACTS FROM ERGODIC THEORY

The perturbation analysis of linear and nonlinear equations leads frequently to problems centering on properties of null spaces of linear operators. In the probabilistic context the linear operators are frequently transition operators or infinitesimal generators. The corresponding study of null spaces is, in effect, their ergodic theory. We shall review now some of the basic notions in ergodic theory which are used later.

Let S be a set and Σ a σ-algebra of subsets and let $P(x, A)$ be a transition probability function,* i.e., a function on $S \times \Sigma \to [0, 1]$ such that $P(\cdot, A)$ is measurable for each $A \in \Sigma$ and $P(x, \cdot)$ is a probability measure for each $x \in S$. Let $B(S)$ and $M(S)$ be the Banach spaces of bounded measurable functions and finite signed measures on S with the sup and total variation norm respectively.

---

* We consider continuous time problems a little further ahead.

The transition function induces an operator T on B and M as follows:

$$Tf(x) = \int P(x, dy)f(y), \qquad f \in B, \tag{3.1}$$

$$\mu T(A) = \int \mu(dx)P(x, A), \qquad \mu \in M. \tag{3.2}$$

We are interested in the behavior of the iterates T^n of the operator T for $n \to \infty$. For $\theta \in [0, 1)$ let R_θ be the resolvent operator*

$$R_\theta = \sum_{n=0}^{\infty} \theta^n T^n = (1 - \theta T)^{-1}. \tag{3.3}$$

We are interested specifically in the existence of the Abelian limit $\lim(1 - \theta)R_\theta$, $\theta \uparrow 1$, which defines a projection operator (projection into the null space of $T - I$) and in addition the existence of solutions to Poisson's equation

$$u - Tu = g \tag{3.4}$$

for given g appropriately restricted.

The simplest result in connection with the above questions is obtained from the following Doeblin's condition:

There exists a reference probability measure ϕ on S and a constant $c > 0$ such that

$$P(x, A) \geqslant c\phi(A). \tag{3.5}$$

Under this condition it follows that if μ and ν are two probability measures

$$\|\mu T^n - \nu T^n\| \leqslant 2(1 - c)^n$$

where the norm is the total variation norm. Thus for any initial measure μ the probabilities μT^n converge geometrically fast to a limit probability $\bar{P}$ which is independent of μ. It is easily seen that $\bar{P}$ is the

* By definition T^0 is the identity operator.

unique invariant probability measure for the transition function $P(x, A)$, i.e.,

$$\bar{P}(A) = \int_S \bar{P}(dx)P(x, A). \tag{3.6}$$

If $\bar{P}$ denotes the operator on $B(S)$ defined by

$$\bar{P}f(x) = \int f(y)\bar{P}(dy) \qquad \text{(constant function)}, \tag{3.7}$$

then clearly $(1 - \theta)R_\theta \to \bar{P}$ as $\theta \uparrow 1$ and $\bar{P}$ is a projection operator.

Moreover, Poisson's equation (3.4) is solvable if and only if

$$\bar{P}g = \int \bar{P}(dy)g(y) = 0, \qquad g \in B(S), \tag{3.8}$$

and the solution is unique up to a constant. The recurrent potential kernel

$$\psi(x, A) = \sum_{n=0}^{\infty} (P^n(x, A) - \bar{P}(A)) \tag{3.9}$$

is well defined and the solution* of (3.4) is

$$u(x) = \int \psi(x, dy)g(y), \qquad g \in B(S). \tag{3.10}$$

In the continuous time case we have a transition function $P(t, x, A)$ satisfying the Chapman–Kolmogorov equation as explained in Section 2. Condition (3.5) is replaced by:

There exists a reference probability measure ϕ on S, a constant $c > 0$ and a $t_0 > 0$ such that

$$P(t_0, x, A) \geqslant c\phi(A). \tag{3.11}$$

Let R_λ be the resolvent operator

$$R_\lambda = \int_0^\infty e^{-\lambda t}T(t)\, dt, \qquad \lambda > 0, \tag{3.12}$$

---

* Up to a constant which will be usually set to zero.

where $T(t)$ is the semigroup on $B(s)$ induced by $P(t, x, A)$. For any $\lambda > 0$ the function $\lambda R_\lambda(x, A)$, defined by

$$\lambda R_\lambda(x, A) = \int_0^\infty \lambda e^{-\lambda t} P(t, x, A)\, dt,$$

the kernel of the resolvent operator, is a transition function and (3.11) implies that

$$\lambda R_\lambda(x, A) \geqslant c\phi(A) e^{-\lambda t_0}. \tag{3.13}$$

Thus, the discrete time result applies to λR_λ. The corresponding invariant measure $\bar{P}$ is independent of $\lambda > 0$ as can be seen from the resolvent identity $R_{\lambda_1} = R_{\lambda_2} + (\lambda_2 - \lambda_1) R_{\lambda_1} R_{\lambda_2}$. This in turn implies that $\bar{P}$ is invariant for the transition function $P(t, x, A)$.

Since

$$\lambda R_\lambda (1 - \theta) \sum_{n=0}^{\infty} \theta^n (\lambda R_\lambda)^n = (1 - \theta) \lambda R_{(1-\theta)\lambda}, \tag{3.14}$$

it follows that $\lim \lambda R_\lambda$ exists as $\lambda \downarrow 0$ and equals the projection operator $\bar{P}$. Moreover the Poisson equation

$$u - \lambda R_\lambda u = R_\lambda g \tag{3.15}$$

has a unique (up to a constant) solution if and only if (3.8) holds. By applying the resolvent identity it follows that u is independent of λ in (3.15) and, in fact, we have (3.10) where

$$\begin{aligned}\psi(x, A) &= R_\lambda \sum_{n=0}^{\infty} [(\lambda R_\lambda)^n - \bar{P}](x, A) \\ &= \int_0^\infty [P(t, x, A) - \bar{P}(A)]\, dt. \end{aligned} \tag{3.16}$$

If Doeblin's condition does not hold, the situation is considerably more complicated but quite a bit is known [12, 13]. Even if one has access to a more general theory, however, if the recurrent potential kernel maps bounded functions to unbounded ones, the nature of the results in the asymptotics changes (cf. Section 13) rather drastically.

Let us suppose that S is a compact metric space with Σ the Borel sets and that the transition function is uniformly Feller, i.e.,

$$\lim_{x \to y} \sup_{A \in \Sigma} |P(x, A) - P(y, A)| = 0. \tag{3.17}$$

This implies that the transition operator T of (3.1) maps the unit ball in $B(S)$ into a compact set, i.e., it is compact. If in addition $P(x, A)$ has no nontrivial invariant sets (relative to some reference measure), then again we are effectively in Doeblin's case. A nontrivial invariant set A is such that $P(x, A) = 1$ for all $x \in A$ and $1 > \phi(A) > 0$ (ϕ the reference measure).

A Feller transition function on a compact state space S has invariant measures, as can be easily deduced by the relative weak compactness of families of probability measures on S. Uniqueness of the invariant measure implies the existence of the projection operator $\bar{P}$. However, in general, uniqueness is not easily shown.

In a locally compact space the question of existence of invariant measures is again not too difficult to settle [14]. However uniqueness is difficult.

We shall conclude this section with a brief comment on the connection of notions of mixing [15] to the above.

When dealing with first order perturbation theory (for example laws of large numbers, etc.) the existence and uniqueness of an invariant measure is enough. Actually the existence of the projection operator $\bar{P}$ is enough.

For second order perturbation theory (central limit theorem for example) it is necessary to have a solvability theory for Poisson's equation which usually is called the Fredholm alternative. Now Doeblin's condition leads to a solvability theory that is very strong: $\psi(x, A)$ exists and maps bounded functions to bounded functions. In the asymptotics less will do. For example: ψ maps bounded functions to $\bar{P}$ integrable ones. Mixing conditions, in one way or another, affect the existence and general behavior of the recurrent potential operator.

The mixing condition

$$\sup_{\|f\| \leq 1} \int \bar{P}(dx) \left| \int [P(x, dy) - \bar{P}(dy)] f(y) \right| = \rho < 1,$$

due to Rosenblatt [15], is considerably weaker than Doeblin's condition and still allows the validity of the central limit theorem. Its connections with recurrent potential theory have not been explored yet, however. For more information about notions of mixing we refer to [16].

4. OSCILLATORY PROBLEMS, AVERAGING, AND LAW OF LARGE NUMBERS FOR STOCHASTIC EQUATIONS

Differential equations whose coefficients are rapidly oscillating functions of time are frequently analyzed by the method of averaging [17]. Since noise fluctuations are to a certain extent similar to oscillations in general (the differences will be elucidated below), averaging provides good motivation for the analysis of many fluctuation phenomena.

We begin with the following frequently used example. Let $x(t)$ be the amplitude of an oscillator with equation of motion*

$$\ddot{x} + \epsilon(\omega^2 x^2 + \dot{x}^2 - 1)\dot{x} + \omega^2 x = 0, \qquad t > 0, \tag{4.1}$$

$$x(0) = x_0, \qquad \dot{x}(0) = \dot{x}_0.$$

The damping term is nonlinear in such a way that we expect a limit cycle to develop when $\epsilon \ll 1$ and time is large. To analyze (4.1) we introduce polar coordinates

$$\begin{aligned} x &= r\cos(\omega t + \theta), \\ \dot{x} &= -\omega r \sin(\omega t + \theta), \end{aligned} \tag{4.2}$$

where $r(t)$ and $\theta(t)$ are functions of time satisfying the system

$$\begin{aligned} \dot{r} &= \epsilon r(1 - \omega^2 r^2)\sin^2(\omega t + \theta), \\ \dot{\theta} &= \frac{\epsilon}{2}(1 - \omega^2 r^2)\sin(2\omega t + 2\theta), \\ r(0) &= r_0, \qquad \theta(0) = \theta_0. \end{aligned} \tag{4.3}$$

Here r_0 and θ_0 are obtained from (4.1) and (4.2).

We note that (4.3) has the following general form:

$$\frac{dx^\epsilon(t)}{dt} = \epsilon F(x^\epsilon(t), t), \qquad x^\epsilon(0) = x, \tag{4.4}$$

* $\dot{x}(t) = dx(t)/dt$.

where $x^\epsilon(t)$ is an n-dimensional vector function of time and $F(x, t)$ is an n-dimensional vector function on $R^n \times [0, \infty)$, periodic in t (in this case, almost periodic more generally). Averaging in its simplest form amounts to this. Let

$$\bar{F}(x) = \lim_{T \uparrow \infty} \frac{1}{T} \int_0^T F(x, s)\, ds \tag{4.5}$$

and let

$$\frac{d\bar{x}(\tau)}{dt} = \bar{F}(\bar{x}(\tau)), \qquad \bar{x}(0) = x. \tag{4.6}$$

Then $x^\epsilon(\tau/\epsilon)$ is well approximated by $\bar{x}(\tau)$ for τ in any finite time interval if ϵ is small.

We first apply this to (4.3). We obtain the system

$$\frac{d\bar{r}(\tau)}{d\tau} = \tfrac{1}{2}\bar{r}(\tau)(1 - \omega^2 \bar{r}^2(\tau)), \qquad \bar{r}(0) = r_0,$$

$$\frac{d\bar{\theta}(\tau)}{d\tau} = 0, \qquad \bar{\theta}(0) = \theta_0. \tag{4.7}$$

Clearly

$$\bar{r}(\tau) = \frac{1}{\omega}\,[1 + e^{-\omega^2 \tau}(\omega^2 r_0^2 - 1)]^{1/2}. \tag{4.8}$$

Thus, the approximation of $r(\tau/\epsilon)$ by $\bar{r}(\tau)$ does capture the qualitative features one may expect from (4.1).

For a more exact statement of the averaging principle we redefine the time scale t so that

$$\frac{dx^\epsilon(t)}{dt} = F\left(x^\epsilon(t), \frac{t}{\epsilon}\right), \qquad x^\epsilon(0) = x, \tag{4.9}$$

with first approximation

$$\frac{d\bar{x}(t)}{dt} = \bar{F}(\bar{x}(t)), \qquad \bar{x}(0) = x, \tag{4.10}$$

$\bar{F}(x)$ being given by (4.5). Suppose that $F(x, t)$ and $\bar{F}(x)$ are smooth functions of x and bounded. Suppose that (4.5) holds in a stronger sense

$$\left| \int_0^t [F(x, s) - \bar{F}(x)]\, ds \right| \leqslant c < \infty,$$

$$\left| \int_0^t [F_x(x, s) - \bar{F}_x(x)]\, ds \right| \leqslant c < \infty, \tag{4.11}$$

for all $x \in R^n$ and $t \geqslant 0$. Then given $T < \infty$ there is an ϵ_0 and a constant c' such that

$$\sup_{0 \leq t \leq T} |x^\epsilon(t) - \bar{x}(t)| < c'\epsilon, \qquad \text{if } \epsilon \leqslant \epsilon_0. \tag{4.12}$$

The proof of this theorem is simple and instructive so we shall give it.

Since (4.11) holds, it suffices to show that

$$w^\epsilon(t) = x^\epsilon(t) - \bar{x}(t) - \epsilon \int_0^{t/\epsilon} [F(x^\epsilon(t), s) - \bar{F}(x^\epsilon(t))]\, ds \tag{4.13}$$

has small norm. But

$$\frac{dw^\epsilon(t)}{dt} = \bar{F}(x^\epsilon(t)) - \bar{F}(\bar{x}(t))$$

$$- \epsilon \int_0^{t/\epsilon} [F_x(x^\epsilon(t), s) - \bar{F}_x(x^\epsilon(t))] F\left(x^\epsilon(t), \frac{t}{\epsilon}\right) ds, \qquad w^\epsilon(0) = 0 \tag{4.14}$$

hence*

$$|w^\epsilon(t)| \leqslant c_1 \int_0^t |w^\epsilon(s)|\, ds + c_2 t\epsilon$$

from which the result follows.

What should be noted in the above simple argument is that the passage from $F(x, t)$ to $\bar{F}(x)$, with x fixed, is the projection (4.5). If $F(x, t)$ is almost periodic uniformly in x then the limit (4.5) exists uniformly in x and uniformly in τ if s is replaced by $s + \tau$ in the integrand. This is all that is necessary for mere convergence of $x^\epsilon(t)$

---

* $c_1, c_2, \ldots$, etc., are constants.

to $\bar{x}(t)$. Hypothesis (4.11) concerns the "recurrent potential" and is strong enough to give the error estimate (4.12).

In the event $\bar{F}(x) \equiv 0$ and it is necessary to go to higher order perturbations, then the recurrent potential will actually appear in the final result, so we must assume that it exists, clearly. Higher order corrections can be constructed even if $\bar{F}(x) \not\equiv 0$ but then the approximations will themselves depend on ϵ. The algorithm for constructing them is well known [17] and we will give it in Section 7.

The literature on problems associated with averaging is huge. Typical questions range from implementation to specific examples, to extensions so that the approximations are valid uniformly in time ($T = \infty$ above), to questions of existence and stability of periodic solutions, etc. We refer to the survey [18] and Hale's book [19] for more information.

We turn now to the stochastic problem which is analogous to (4.9). To make the connections with Section 3 more transparent we shall deal with the following model.

Let $\{y(t), t \geqslant 0\}$ be a Markov process* on a state space (S, Σ) and let $P(t, y, A)$ be its transition function. Assume that it is ergodic with invariant measure $\bar{P}(A)$ and such that

$$\lim_{\lambda \downarrow 0} \lambda R_\lambda f(y) = \lim_{\lambda \downarrow 0} \lambda \int_0^\infty e^{-\lambda t} T(t) f(y)\, dt$$

$$= \bar{P} f = \int \bar{P}(dy) f(y), \tag{4.15}$$

uniformly in y for $f \in B(S)$. Here $T(t)$ is the semigroup associated with $P(t, y, A)$.

Let $F(x, y)$ be a bounded measurable n-vector function on $R^n \times S$, with bounded x derivatives. Consider the system

$$\frac{dx^\epsilon(t)}{dt} = F(x^\epsilon(t), y^\epsilon(t)), \qquad x^\epsilon(0) = x, \tag{4.16a}$$

where we have defined

$$y^\epsilon(t) = y(t/\epsilon). \tag{4.16b}$$

---

* Thus, $y(t) = y(t, \omega)$, $\omega \in \Omega$, $(\Omega, \mathscr{F}, P)$ a probability space. Similarly in (4.16) $x^\epsilon(t) = x^\epsilon(t, \omega)$. We shall not indicate the ω explicitly in the sequel.

When S is the real line, $F(x, y)$ is almost periodic in y and the transition function corresponds to deterministic motion at unit speed, we recover (4.9).

It is natural that $\bar{F}(x)$ in (4.5) be replaced by

$$\bar{F}(x) = \int \bar{P}(dy)F(x, y) \tag{4.17}$$

and that

$$\frac{d\bar{x}(t)}{dt} = \bar{F}(\bar{x}(t)), \qquad \bar{x}(0) = x \tag{4.18}$$

should in some sense approximate (4.16).

A simple result along these lines is that if the recurrent potential exists (analog of (4.11)), i.e., (3.16) is well defined, then

$$\lim_{\epsilon \downarrow 0} P_{x,y}\left\{ \sup_{0 \leq t \leq T} |x^\epsilon(t) - \bar{x}(t)| > \delta \right\} = 0, \qquad T < \infty, \tag{4.19}$$

for any $\delta > 0$ and uniformly in $(x, y) \in R^n \times S$. We may call this the law of large numbers.

The terminology is justified because for $n = 1$ and $F(x, y) = \chi_A(y)$ independent of x with $A \in \Sigma$, we have

$$\begin{aligned} x^\epsilon(t) &= x + \int_0^t \chi_A(y^\epsilon(s))\, ds \\ &= x + \int_0^t \chi_A(y(s/\epsilon))\, ds \\ &= x + \epsilon \int_0^{t/\epsilon} \chi_A(y(s))\, ds, \end{aligned}$$

which converges with probability one in fact (by the ergodic theorem) to $x + t\bar{P}(A)$ as $\epsilon \to 0$.

The proof is similar to the one for averaging. First we define

$$\psi_F(x, y) = \int_0^\infty \int [P(t, y, dz) - \bar{P}(dz)]F(x, z)\, dt \tag{4.20}$$

so that

$$-A\psi_F = F, \tag{4.21}$$

A being the infinitesimal generator of $[y(t), t \geqslant 0]$ (cf. (3.15), (3.16)). Then we note that $(x^\epsilon(t), y^\epsilon(t))$ is Markovian on $R^n \times S$ with generator

$$\frac{1}{\epsilon} A + F(x, y) \cdot \frac{\partial}{\partial x}$$

with the dot denoting the inner product and $\partial/\partial x$ the gradient operator. From the discussion of Section 2 we know that

$$x^\epsilon(t) + \epsilon\psi_F(x^\epsilon(t), y^\epsilon(t)) - \int_0^t \left(\frac{1}{\epsilon} A + F \cdot \frac{\partial}{\partial x}\right)(x + \epsilon\psi_F)(x^\epsilon(s), y^\epsilon(s))\, ds$$

is a martingale. In fact* the expression

$$\begin{aligned} x^\epsilon(t) &+ \epsilon\psi_F(x^\epsilon(t), y^\epsilon(t)) - x - \epsilon\psi(x, y) \\ &- \int_0^t \left(\frac{1}{\epsilon} A + F \cdot \frac{\partial}{\partial x}\right)(x + \epsilon\psi_F)(x^\epsilon(s), y^\epsilon(s))\, ds \equiv M^\epsilon(t) \end{aligned} \tag{4.22}$$

is a zero mean martingale. Using (4.21) and the fact that $A1 = 0$ (A is a Markovian generator), we find that

$$\begin{aligned} x^\epsilon(t) - x - \int_0^t \bar{F}(x^\epsilon(s))\, ds \\ &= M^\epsilon(t) + \epsilon(\psi_F(x, y) - \psi_F(x^\epsilon(t), y^\epsilon(t))) \\ &\quad + \epsilon \int_0^t F \cdot \frac{\partial}{\partial x} \psi_F(x^\epsilon(s), y^\epsilon(s))\, ds. \end{aligned}$$

Subtracting the integrated version of (4.18) from this yields

$$\begin{aligned} x^\epsilon(t) - \bar{x}(t) &= \int_0^t [F(x^\epsilon(s)) - \bar{F}(\bar{x}(s))]\, ds + M^\epsilon(t) \\ &\quad + \epsilon(\psi_F(x, y) - \psi_F(x^\epsilon(t), y^\epsilon(t))) \\ &\quad + \epsilon \int_0^t F \cdot \frac{\partial \psi_F}{\partial x}(x^\epsilon(s), y^\epsilon(s))\, ds. \end{aligned} \tag{4.23}$$

This is the analog of (4.13) and (4.14) (combined).

---

* The notation in the integral means that the operator acts on the function shown and the result is evaluated at the argument shown.

To get the result (4.19) and (4.23) is now an easy matter. It is roughly the same process as in the deterministic case except now the martingale term $M^\epsilon(t)$ must also be estimated. Since the second moments of $M^\epsilon(t)$ are proportional to ϵ an application of Kolmogorov's inequality is all that is necessary.

As a final note we should point out that most of the extensions and side issues associated with averaging make sense for the stochastic problem also. However, what is more interesting is that in the stochastic case one can ask many additional questions that have no real analogs in the deterministic case; for example, the fluctuation theory of $x^\epsilon(t)$ about $\bar{x}(t)$ (properly scaled) as discussed in Section 7. The second order perturbation theory associated with (4.16), when $\bar{F}(x) \equiv 0$ in (4.17), is discussed in Section 7. Further information regarding averaging can be found in the papers of Kurtz [20], for example, [21], and the next section. Some extensions and applications of averaging can be found in [22] and [23], as well as in [24].

5. SPATIAL AVERAGING AND HOMOGENIZATION

We now consider an example treated by Freidlin [25], Khasminskii [26], and in [27]. We pose the problem in a deterministic way at first.

Let $(a_{ij}(y))$ and $(b_j(y))$ be a symmetric positive definite $n \times n$ matrix and n-vector function of $y \in S$, respectively, where S is the unit torus on R^n, assumed smooth. Consider the diffusion equation

$$\frac{\partial u^\epsilon(t,x)}{\partial t} = \frac{1}{2}\sum_{i,j=1}^{n} a_{ij}\left(\frac{x}{\epsilon}\right)\frac{\partial^2 u^\epsilon(t,x)}{\partial x_i \partial x_j} + \sum_{j=1}^{n} b_j\left(\frac{x}{\epsilon}\right)\frac{\partial u^\epsilon(t,x)}{\partial x_j}, \quad (5.1)$$

$$t > 0, \qquad x \in \mathscr{D},$$

$$u^\epsilon(0,x) = f(x), \qquad u^\epsilon(t,x) = 0, \qquad x \in \partial\mathscr{D}, \qquad t > 0.$$

Here $\mathscr{D}$ is a bounded open set in R^n and $\partial\mathscr{D}$ is its assumed smooth boundary.

The question is: how does u^ϵ behave as $\epsilon \to 0$ and the coefficients, being periodic, oscillate more and more rapidly?

Let us assume that the operator

$$A = \frac{1}{2}\sum_{i,j=1}^{n} a_{ij}(y)\frac{\partial^2}{\partial y_i \partial y_j} \quad (5.2)$$

is uniformly elliptic on S and let $[y(t), t \geqslant 0]$ be the diffusion process* it generates on S. The uniform ellipticity implies clearly that $y(t)$ is ergodic and, by smoothness of $a_{ij}(y)$, the invariant measure has a density $\bar{p}(y)$ relative to Lebesgue measure on S. Furthermore, the recurrent potential kernel exists since Doeblin's theory applies via the strong maximum principle.

Define

$$\bar{a}_{ij} = \int_S a_{ij}(y)\bar{p}(y)\, dy,$$

$$\bar{b}_{ij} = \int_S b_j(y)\bar{p}(y)\, dy, \qquad i, j = 1, 2, \ldots, n. \tag{5.3}$$

Then $u^\epsilon(t, x)$ converges uniformly† in $x \in \mathscr{D}$, $0 \leqslant t \leqslant T$ to the solution $\bar{u}(t, x)$ of

$$\frac{\partial \bar{u}(t, x)}{\partial t} = \frac{1}{2} \sum_{i,j=1}^{n} \bar{a}_{ij} \frac{\partial^2 \bar{u}(t, x)}{\partial x_i \partial x_j} + \sum_{i=1}^{n} \bar{b}_j \frac{\partial \bar{u}(t, x)}{\partial x_j}, \tag{5.4}$$

$$u(0, x) = f(x), \qquad \bar{u}(t, x) = 0, \qquad t > 0, \qquad x \in \partial\mathscr{D}.$$

Thus the limit problem has constant coefficients which are appropriate averages of the original ones. This explains the terminology.

If $a_{ij} = a_{ij}(x, y)$, $b_j = b_j(x, y)$, $x \in \mathscr{D}$, $y \in S$, then x is carried along as a parameter in (5.2) and (5.3) and $\bar{a}_{ij}$, $\bar{b}_j$ depend now on x.

In terms of stochastic differential equations, the process generated by the operator on the right side of (5.1) satisfies‡

$$dx^\epsilon(t) = b\left(\frac{x^\epsilon(t)}{\epsilon}\right) dt + \sigma\left(\frac{x^\epsilon(t)}{\epsilon}\right) dw(t), \qquad x^\epsilon(0) = x, \tag{5.5}$$

where $\sigma(y)$ is the symmetric (smooth, since a is smooth and uniformly elliptic) square root of $a(y)$ and $w(t)$ is the standard n-dimensional Brownian motion. The convergence corresponds to weak convergence on $C([0, T]; R^n)$ of the corresponding measures. The limit measure is, essentially, Brownian motion.

---

* Thus $y(t) = y(t, \omega)$, $\omega \in \Omega$, $(\Omega, \mathscr{F}, P)$ a probability space.

† We also assume that $f(x) \equiv 0$ outside a compact subset of $\mathscr{D}$ and is smooth.

‡ The argument $\omega \in \Omega$, $(\Omega, \mathscr{F}, P)$ a probability space, is omitted.

We outline a proof for (5.1)–(5.4) without too many details and without its probabilistic ramifications.

First, and this is a useful device, we let $y = x/\epsilon$ and construct a function $v^\epsilon(t, x, y)$ such that

$$\frac{\partial v^\epsilon(t, x, y)}{\partial t}$$

$$= \frac{1}{\epsilon^2}\left[\frac{1}{2}\sum_{i,j=1}^{n} a_{ij}(y)\frac{\partial^2 v^\epsilon(t, x, y)}{\partial y_i \partial y_j}\right]$$

$$+ \frac{1}{\epsilon}\left[\sum_{i,j=1}^{n} a_{ij}(y)\frac{\partial^2 v^\epsilon(t, x, y)}{\partial y_i \partial x_j} + \sum_{j=1}^{n} b_j(y)\frac{\partial v^\epsilon(t, x, y)}{\partial y_j}\right]$$

$$+ \left[\frac{1}{2}\sum_{i,j=1}^{n} a_{ij}(y)\frac{\partial^2 v^\epsilon(t, x, y)}{\partial x_i \partial x_j} + \sum_{j=1}^{n} b_i(y)\frac{\partial v^\epsilon(t, x, y)}{\partial x_j}\right] \tag{5.6}$$

this step being formal so that we do not specify boundary conditions, etc. Note if v^ϵ is appropriately constructed $u^\epsilon(t, x) = v^\epsilon(t, x, x/\epsilon)$ and this is the motivation for this step.

Next we look for a power series expansion of $v^\epsilon(t, x, y)$

$$v^\epsilon = v_0 + \epsilon v_1 + \epsilon^2 v_2 + \cdots. \tag{5.7}$$

Inserting (5.7) into (5.6) and equating coefficients of powers of ϵ we obtain

$$\frac{1}{2}\sum_{i,j=1}^{n} a_{ij}(y)\frac{\partial^2 v_0}{\partial y_i \partial y_j} = 0, \tag{5.8}$$

$$\frac{1}{2}\sum_{i,j=1}^{n} a_{ij}(y)\frac{\partial^2 v_1}{\partial y_i \partial y_j} + \sum_{i,j=1}^{n} a_{ij}(y)\frac{\partial^2 v_0}{\partial y_i \partial x_j} + \sum_{j=1}^{n} b_j(y)\frac{\partial v_0}{\partial y_j} = 0, \tag{5.9}$$

$$\frac{1}{2}\sum_{i,j=1}^{n} a_{ij}(y)\frac{\partial^2 v_2}{\partial y_i \partial y_j} + \sum_{i,j=1}^{n} a_{ij}(y)\frac{\partial^2 v_1}{\partial y_i \partial x_j} + \sum_{j=1}^{n} b_j(y)\frac{\partial v_1}{\partial y_j}$$

$$+ \frac{1}{2}\sum_{i,j=1}^{n} a_{ij}(y)\frac{\partial^2 v_0}{\partial x_i \partial x_j} + \sum_{j=1}^{n} b_j(y)\frac{\partial v_0}{\partial x_j} - \frac{\partial v_0}{\partial t} = 0. \tag{5.10}$$

From (5.8) and the ergodicity of A in (5.2) we conclude that $v_0 = v_0(t, x)$. We may thus take $v_1 \equiv 0$ in (5.9). Thus (5.10) takes the form

$$Av_2 + \frac{1}{2} \sum_{i,j=1}^{n} a_{ij}(y) \frac{\partial^2 v_0(t, x)}{\partial x_i \partial x_j} + \sum_{j=1}^{n} b_j(y) \frac{\partial v_0(t, x)}{\partial x_j} - \frac{\partial v_0}{\partial t} = 0. \tag{5.11}$$

In order that (5.11) have a bounded solution the inhomogeneous term must integrate to zero relative to the invariant measure $\bar{p}(y)\, dy$; (5.11) is just Poisson's equation (3.15)–(3.16). Applying this solvability condition we find that $v_0(t, x)$ must satisfy (5.4) (and now we insert initial and boundary condition).

After this formal step the proof goes as follows. Let $\bar{u}(t, x)$ solve (5.4) and be smooth. Let v_2 be any bounded solution of (5.11). Then if

$$w^\epsilon(t, x) = u^\epsilon(t, x) - \bar{u}(t, x) - \epsilon^2 v_2\left(t, x, \frac{x}{\epsilon}\right),$$

we find that by construction

$$\left[\frac{\partial}{\partial t} - \frac{1}{2} \sum_{i,j=1}^{n} a_{ij}\left(\frac{x}{\epsilon}\right) \frac{\partial^2}{\partial x_i \partial x_j} - \sum_{j=1}^{n} b_j\left(\frac{x}{\epsilon}\right) \frac{\partial}{\partial x_j}\right] w^\epsilon(t, x) = O(\epsilon),$$

$$w^\epsilon(0, x) = O(\epsilon^2), \qquad x \in \mathscr{D},$$

$$w^\epsilon(t, x) = O(\epsilon^2), \qquad x \in \partial\mathscr{D}, \qquad t > 0.$$

The maximum principle now yields the desired result.

Homogenization is a subject with many ramifications and a surprisingly diverse field of applications. A systematic treatment will be given in a forthcoming book [28]. We also refer to the papers of Babuška [29] for further results and applications.

6. THE GAUSS-MARKOV APPROXIMATION AND APPLICATIONS

We continue the analysis of (4.16) under the assumptions stated there and with (4.19) established. To analyze fluctuations* set

$$x^\epsilon(t) = \bar{x}(t) + \sqrt{\epsilon}\, z^\epsilon(t). \tag{6.1}$$

* Again the $\omega \in \Omega$, $(\Omega, \mathscr{F}, P)$ a probability space, is omitted.

Then,

$$\frac{dz^\epsilon(t)}{dt} = \frac{1}{\sqrt{\epsilon}}[F(\bar{x}(t) + \sqrt{\epsilon}z^\epsilon(t), y^\epsilon(t)) - F(\bar{x}(t))] \tag{6.2}$$

$$z^\epsilon(0) = 0.$$

The problem is to find the asymptotic limit of the fluctuation process $z^\epsilon(t)$ as $\epsilon \to 0$.

Let us first compute formally. Expanding the right-hand side of (6.2) we obtain

$$\frac{dz^\epsilon(t)}{dt} = \frac{1}{\sqrt{\epsilon}}[F(\bar{x}(t), y^\epsilon(t)) - \bar{F}(\bar{x}(t))] + \frac{\partial F(\bar{x}(t), y^\epsilon(t))}{\partial x} z^\epsilon(t) + O(\sqrt{\epsilon}). \tag{6.3}$$

Hence

$$z^\epsilon(t) = \frac{1}{\sqrt{\epsilon}}\int_0^t [F(\bar{x}(s), y^\epsilon(s)) - \bar{F}(\bar{x}(s))]\,ds + \int_0^t \frac{\partial F(\bar{x}(s), y^\epsilon(s))}{\partial x} z^\epsilon(s)\,ds + O(\sqrt{\epsilon}). \tag{6.4}$$

Let us look first at the first term on the right-hand side of (6.4). Recalling that $y^\epsilon(t) = y(t/\epsilon)$ we can change variables to obtain

$$\sqrt{\epsilon}\int_0^{t/\epsilon} [F(\bar{x}(\epsilon s), y(s)) - \bar{F}(\bar{x}(\epsilon s))]\,ds.$$

If it were not for the $\bar{x}(t)$ dependence the problem would amount to the central limit theorem for a function of an ergodic Markov process. If we let

$$a_{ij}(x) = \int \bar{P}(dy)\int \psi(y, dy')(F_i(x, y) - \bar{F}_i(y)) \cdot (F_j(x, y') - \bar{F}_j(y')), \qquad i, j = 1, \ldots, n, \tag{6.5}$$

where $\psi(y, A)$ is the recurrent potential kernel (3.16), and $(\sigma_{ij}(x))$ is the symmetric square root of (a_{ij}), then the formal limit of $z^\epsilon(t)$ satisfies

$$z(t) = \int_0^t \sigma(\bar{x}(s))\, dw(s) + \int_0^t \frac{\partial \bar{F}(\bar{x}(s))}{\partial x} z(s)\, ds, \tag{6.6}$$

where $w(t)$ is the standard n-dimensional Brownian motion.

Thus $z(t)$ is a time-inhomogeneous Gaussian process with independent increments satisfying the linear stochastic differential equation (6.6) where $\bar{x}(t)$ satisfies (4.17).

Let $U(t, s)$ be the fundamental solution of the linear variational equation to (4.18)

$$\frac{dU(t, s)}{dt} = \frac{\partial \bar{F}(\bar{x}(t))}{\partial x} U(t, s), \qquad t > s, \tag{6.7}$$

$$U(s, s) = I \text{ (identity)}.$$

Let

$$v(t) = \int_0^t \sigma(\bar{x}(s))\, dw(s)$$

which is a Gaussian process with independent increments. Then,

$$z(t) = \int_0^t U(t, s)\, dv(s) \tag{6.8}$$

and so

$$E\{z(t) z^T(t)\} = \int_0^t ds\, U(t, s) a(\bar{x}(s)) U^T(t, s) \tag{6.9}$$

where T denotes transpose.

The above description points to the fact that the fluctuation process has a nice and relatively simple asymptotic behavior. We should stress that (i) the approximation (which can be easily proved) is valid only on finite but arbitrary time intervals and this limits its potential usefulness and (ii) the approximation is a fairly crude one and makes sense in great generality (in infinite dimensional, for example, partial differential equations, problems). For a proof we refer to [3] and [30] as well as [31–33].

A simple proof can be given along the following lines. We consider jointly $(z^\epsilon(t), \bar{x}(t), y^\epsilon(t))$ which is a Markov process on $R^n \times R^n \times S$

with infinitesimal generator*

$$\mathscr{L}^\epsilon = \frac{1}{\epsilon} A + \frac{1}{\sqrt{\epsilon}} (F(\bar{x} + \sqrt{\epsilon} z, y) - \bar{F}(\bar{x}) \cdot \frac{\partial}{\partial z} + \bar{F}(\bar{x}) \cdot \frac{\partial}{\partial x}. \quad (6.10)$$

Evidently $\bar{x}(t)$ is deterministic and decouples from $(z^\epsilon(t), y^\epsilon(t))$ but it is carried along to make things time homogeneous. Now note that $\mathscr{L}^\epsilon$ has the form

$$\mathscr{L}^\epsilon = \frac{1}{\epsilon} \mathscr{L}_1 + \frac{1}{\sqrt{\epsilon}} \mathscr{L}_2 + \mathscr{L}_3 + O(\sqrt{\epsilon}) \quad (6.11)$$

where $\mathscr{L}_1 = A$, $\mathscr{L}_2 = (F(\bar{x}, y) - \bar{F}(\bar{x})) \cdot \partial/\partial x$ and

$$\mathscr{L}_3 = \bar{F}(\bar{x}) \cdot \frac{\partial}{\partial x} + \frac{\partial F(\bar{x}, y)}{\partial x} z \cdot \frac{\partial}{\partial z}.$$

If $T^\epsilon(t)$ denotes the semigroup generated by $\mathscr{L}^\epsilon$ on $C(R^n \times R^n \times S)$ and if

$$\bar{\mathscr{L}} = \sum_{i,j=1}^n a_{ij}(\bar{x}) \frac{\partial^2}{\partial x_i \partial z_j} + \sum_{i,j=1}^n \frac{\partial \bar{F}_i(\bar{x})}{\partial \bar{x}_j} z_j \frac{\partial}{\partial z_i} + \sum_{i=1}^n \bar{F}_i(\bar{x}) \frac{\partial}{\partial \bar{x}_i} \quad (6.12)$$

is the generator of the limit process $(z(t), \bar{x}(t))$ with semigroup $T(t)$ on $C(R^n \times R^n)$, then we must show that $T^\epsilon(t) f \underset{\epsilon \downarrow 0}{\rightarrow} T(t) f$ for each $f \in C(R^n \times R^n)$ and smooth. Compactness of the corresponding measures can also be established without difficulty. We are therefore in the situation described at the end of Section 2.

Because of the singular form of $\mathscr{L}^\epsilon$ in (6.11) the generators do not converge when acting on fixed functions. Following Kurtz [11], for given $f(z, \bar{x})$ smooth we define $f_1(z, \bar{x}, y)$ and $f_2(z, \bar{x}, y)$ as solutions of the following Poisson's equations

$$A f_1 + (F(\bar{x}, y) - \bar{F}(\bar{x})) \cdot \frac{\partial f(z, \bar{x})}{\partial z} = 0,$$

$$A f_2 + (F(\bar{x}, y) - \bar{F}(\bar{x})) \cdot \frac{\partial f_1(z, \bar{x}, y)}{\partial z} + \frac{\partial F(\bar{x}, y)}{\partial \bar{x}} z \cdot \frac{\partial f(z, \bar{x})}{\partial z}$$

$$+ \bar{F}(\bar{x}) \cdot \frac{\partial f(z, \bar{x})}{\partial \bar{x}} - \bar{\mathscr{L}} f(z, \bar{x}) = 0. \quad (6.13)$$

---

* A is the infinitesimal generator of $y(t)$.

Note that by definition of $\bar{F}$ in (4.17), and of $\bar{\mathscr{L}}$ in (6.12) the solvability condition for the equations (6.13), under the strong ergodicity assumptions we have here, f_1 and f_2 are well defined. Now set

$$f^{\epsilon}(z, \bar{x}, y) = f(z, \bar{x}) + \sqrt{\epsilon} f_1(z, \bar{x}, y) + \epsilon f_2(z, \bar{x}, y). \quad (6.14)$$

By construction

$$\mathscr{L}^{\epsilon} f^{\epsilon} = \bar{\mathscr{L}} f + O(\sqrt{\epsilon})$$

and since $f^{\epsilon} \to f$ clearly the result follows as was explained at the end of Section 2.

It should be noted that the above construction of f_1 and f_2 can be motivated simply by inserting power series and equating coefficients as in (5.7)–(5.11). We refer to the above as a second order result because one has to carry perturbation theory one order further than in Section 4 which is the first order result.

7. DIFFUSION APPROXIMATIONS

We now consider the problem

$$\frac{dx^{\epsilon}(t)}{dt} = \frac{1}{\epsilon} F(x^{\epsilon}(t), y^{\epsilon}(t)), \qquad x^{\epsilon}(0) = x,$$

$$y^{\epsilon}(t) = y(t/\epsilon^2), \quad (7.1)$$

where $F(x, y)$ is a bounded measurable n-vector function on $R^n \times S$ with bounded x-derivatives and $y(t)$ is an ergodic Markov process on a state space (S, Σ) with transition function $P(t, y, B)$, infinitesimal generator A, invariant measure $\bar{P}(B)$ and recurrent potential $\psi(x, B)$ (cf. (3.16)). We assume in this section that

$$\int_S F(x, y)\bar{P}(dy) \equiv 0 \quad (7.2)$$

and we have changed the scaling in (7.1) as compared to (4.16), (4.17) in order to deal with integral powers of ϵ and with the limit being $O(1)$.

From the result (4.19) it follows that on time intervals of the form $0 \leqslant t \leqslant \epsilon C$, C is a constant, the process $x^{\epsilon}(t)$ does not move much from its initial value. On the larger time intervals $0 \leqslant t \leqslant T < \infty$ we shall show that $x^\epsilon(t)$ behaves like a diffusion Markov process.

First we consider $(x^\epsilon(t), y^\epsilon(t))$ jointly as a Markov process on $R^n \times S$. The infinitesimal generator of this process is easily seen to have the following form on smooth functions

$$\mathscr{L}^\epsilon = \frac{1}{\epsilon^2} A + \frac{1}{\epsilon} F(x, y) \cdot \frac{\partial}{\partial x}. \tag{7.3}$$

The semigroup $T^\epsilon(t)$ is given by

$$T^\epsilon(t) f(x, y) = E_{x,y}\{f(x^\epsilon(t), y^\epsilon(t))\}, \tag{7.4}$$

and the process $x^\epsilon(t)$ ($y^\epsilon(t)$ is given) is constructed pathwise.

To see how $T^\epsilon(t)f$ behaves when $\epsilon \downarrow 0$ and $f = f(x)$ (we look at the $x^\epsilon(t)$ process only) and we set

$$u^\epsilon(t, x, y) = T^\epsilon(t) f(x, y), \tag{7.5}$$

with $f(x)$ smooth, and thus

$$\frac{\partial u^\epsilon(t, x, y)}{\partial t} = \frac{1}{\epsilon^2} A u^\epsilon(t, x, y) + \frac{1}{\epsilon} F(x, y) \cdot \frac{\partial u^\epsilon(t, x, y)}{\partial x} \tag{7.6}$$

$$u^\epsilon(0, x, y) = f(x).$$

We analyze (7.6) by the usual methods of asymptotic expansions. Put

$$u^\epsilon = u_0 + \epsilon u_1 + \epsilon^2 u_2 + \cdots, \tag{7.7}$$

insert this in (7.6), and equate coefficients of equal powers of ϵ to obtain the following sequence of problems

$$A u_0 = 0 \tag{7.8}$$

$$A u_1 + F \cdot \frac{\partial u_0}{\partial x} = 0 \tag{7.9}$$

$$A u_2 + F \cdot \frac{\partial u_1}{\partial x} - \frac{\partial u_0}{\partial t} = 0, \text{ etc.} \tag{7.10}$$

Since A acts on u_0 as a function of y alone and A is ergodic ($A1 = 0$ and 1 is the only such solution), $u_0 = u_0(t, x)$ but it is not yet determined further.

Now we look at (7.9). It is a Poisson's equation in S space with x and t being parameters. Since A is ergodic (7.9) will have a solution provided the solvability condition

$$\int \bar{P}(dy) F \cdot \frac{\partial u_0}{\partial x} = 0 \tag{7.11}$$

holds. But this is (7.2) and therefore u_1 exists and is given by

$$u_1(t, x, y) = \int_S \psi(y, dz) F(x, z) \cdot \frac{\partial u_0(t, x)}{\partial x}. \tag{7.12}$$

Next we consider (7.10). The solvability condition for it is

$$\int \bar{P}(dy) \left[F(x, y) \cdot \frac{\partial u_1(t, x, y)}{\partial x} - \frac{\partial u_0(t, x)}{\partial t} \right] = 0. \tag{7.13}$$

Using (7.12) in (7.13) yields the following diffusion equation for $u_0(t, x)$.

$$\frac{\partial u_0(t, x)}{\partial t} = \int \bar{P}(dy) \int \psi(y, dz) \sum_{i,j=1}^{n} F_i(x, y) \frac{\partial}{\partial x_i} \cdot \left(F_j(x, z) \frac{\partial u_0(t, x)}{\partial x_j} \right). \tag{7.14}$$

The diffusion and drift coefficients are given by

$$a_{ij}(x) = \int \bar{P}(dy) \int \psi(y, dz) F_i(x, y) F_j(x, z), \tag{7.15}$$

$$b_j(x) = \int \bar{P}(dy) \int \psi(y, dz) \sum_{i=1}^{n} F_i(x, y) \frac{\partial F_j(x, z)}{\partial x_i}, \tag{7.16}$$

and hence (7.14) becomes

$$\frac{\partial u_0(t, x)}{\partial t} = \bar{\mathscr{L}} u_0 \equiv \sum_{i,j=1}^{n} a_{ij}(x) \frac{\partial^2 u_0(t, x)}{\partial x_i \partial x_j} + \sum_{j=1}^{n} b_j(x) \frac{\partial u_0(t, x)}{\partial x_j}, \tag{7.17}$$

$$u_0(t, x) = f(x).$$

Let $T(t)f(x) = u_0(t, x)$, $T(t)$ being the semigroup of the diffusion process $x(t)$. To show that for $f = f(x)$ smooth $T^\epsilon(t)f \to T(t)f$, $0 \leqslant t \leqslant T < \infty$, as $\epsilon \downarrow 0$ is now an easy matter. One way is to exploit the remarks at the end of Section 2 with the choice

$$\begin{aligned} f^\epsilon(x, y) &= f(x) + \epsilon f_1(x, y) + \epsilon^2 f_2(x, y), \\ f_1(x, y) &= \int_S \psi(y, dz) F(x, z) \frac{\partial f(x)}{\partial x}, \\ f_2(x, y) &= \int_S \psi(y, dz) \left[F(x, z) \cdot \frac{\partial f_1(x, z)}{\partial x} - \bar{\mathscr{L}} f(x) \right]. \end{aligned} \tag{7.18}$$

Another way is to simply insert into (7.6) the expression $u^\epsilon - u_0 - \epsilon u_1 - \epsilon^2 u_2$ and use the maximum principle to show that it is small (as was done in Section 5).

From the point of view of applications the most important thing is the formulas (7.15) and (7.16). Let $y(t)$, $-\infty < t < \infty$, be the stationary Markov process on S with transition function $P(t, y, B)$ and invariant measure $\bar{P}(B)$. First note that, in general, a_{ij} is not symmetric in (7.15) but only the symmetric part of a_{ij} enters in (7.17). Now the symmetric part of a_{ij} can also be written in the form

$$\int_{-\infty}^{\infty} dt\, E\{F_i(x, y(0)) F_j(x, y(t))\}, \qquad i, j = 1, 2, \ldots, n, \tag{7.19}$$

where $E\{\cdot\}$ is expectation relative to the stationary process $y(t)$. It is clear that unless the correlations of $F(x, y(t))$ (x is a fixed parameter) die out with increasing separation, (7.19) will not exist. In the case of averaging, where

$$\bar{P} = \lim_{T \uparrow \infty} \frac{1}{T} \int_0^T \cdot\, dt, \qquad \text{(cf. (4.5))}$$

it is clear that the symmetric part of a_{ij} will always come out to be zero: one cannot get diffusive behavior from deterministic oscillations. To get diffusive behavior, the spectrum of the stationary process $y(t)$ (or $F(x, y(t))$, x-fixed) must be continuous at the origin. This is a minimum requirement, a necessary condition.

On the other hand, the drift coefficient (7.16), which has also the form

$$b_j(x) = \int_0^\infty dt\, E\left\{\sum_{i=1}^n F_i(x, y(0)) \frac{\partial F_j(x, y(t))}{\partial x_i}\right\}, \tag{7.19'}$$

can be nonzero even for "deterministic" stationary processes; for example $F(x, y(t))$ almost periodic in t in which case (7.19′) is actually given by a Cesaro limit. In the almost periodic case we recover the second order averaging formulas [17].

The above can be extended a bit to deal with processes $y(t)$ that are ergodic in a strong sense (0 is an isolated point of the spectrum of the generator A) but they can still have isolated point eigenvalues on the imaginary axis (corresponding to cyclically moving sets in the ergodic decomposition of the state space S). The simplest instance of this is when Q is an $n \times n$ skew-symmetric matrix with isolated eigenvalues and (7.1) is replaced by

$$\frac{dx^\epsilon(t)}{dt} = \frac{1}{\epsilon^2} Qx^\epsilon(t) + \frac{1}{\epsilon} F(x^\epsilon(t), y^\epsilon(t)), \qquad x^\epsilon(0) = x. \tag{7.20}$$

Then, if

$$\tilde{x}^\epsilon(t) = e^{-Qt/\epsilon^2} x^\epsilon(t), \qquad \tau^\epsilon(t) = \tau + t/\epsilon^2 \tag{7.21}$$

the process $(x^\epsilon(t), y^\epsilon(t), \tau^\epsilon(t))$ on $(R^n \times S \times R^1)$ is a Markov process with generator

$$\mathscr{L}^\epsilon = \frac{1}{\epsilon^2}\left(A + \frac{\partial}{\partial \tau}\right) + \frac{1}{\epsilon} e^{-Q\tau} F(e^{Q\tau}\tilde{x}, y) \cdot \frac{\partial}{\partial \tilde{x}}. \tag{7.22}$$

Now the "y process" is defined on $S \times R^1$ with generator $A + \partial/\partial\tau$. Since A and $\partial/\partial\tau$ commute and $\partial/\partial\tau$ (acting on almost periodic functions if $F(x, y)$ is analytic in x, for example) has discrete pure imaginary spectrum we are in the situation described above.

Note that the transformation (7.21) picks out the envelope (slowly varying part) of the process $x^\epsilon(t)$. The operator $\bar{\mathscr{L}}$ has now the form

$$\bar{\mathscr{L}} f(x) = \lim_{T\uparrow\infty} \frac{1}{T}\int_0^T dt \int_0^\infty ds \int \bar{P}(dy)[P(s, y, dz) - \bar{P}(dz)]$$

$$e^{-Qt} F(e^{Qt}x, y) \cdot \frac{\partial}{\partial x}\left(e^{-Q(t+s)} F(e^{Q(t+s)}x, z) \cdot \frac{\partial f(x)}{\partial x}\right) \tag{7.23}$$

and the result is that $x^\epsilon(t)$ converges (weakly) as $\epsilon \to 0$ to the diffusion with generator $\bar{\mathscr{L}}$ given by (7.23). An application is given in Section 15.

References for the above are [34–37] with applications in [38]. Operator treatment is given in [37], [40], with diverse applications, and in [39], [41].

We note finally that if instead of (7.20) we have

$$\frac{dx^\epsilon(t)}{dt} = \frac{1}{\epsilon} Qx^\epsilon(t) + \frac{1}{\epsilon} F(x^\epsilon(t), y^\epsilon(t)), \qquad x^\epsilon(0) = x, \tag{7.24}$$

and if

$$\tilde{x}^\epsilon(t) = e^{-Qt/\epsilon} x^\epsilon(t), \tag{7.25}$$

then $\tilde{x}^\epsilon(t)$ converges (weakly) as $\epsilon \to 0$ to the diffusion process generated by

$$\bar{\mathscr{L}} f(x) = \lim_{T \uparrow \infty} \frac{1}{T} \int_0^T dt \int \bar{P}(dy) \int \psi(y, dz)$$

$$\cdot\, e^{-Qt} F(e^{Qt} x, y) \cdot \frac{\partial}{\partial x} \left(e^{-Qt} F(e^{Qt}, z) \cdot \frac{\partial f(x)}{\partial x} \right). \tag{7.26}$$

We shall use this in Section 16.

8. DIFFUSION APPROXIMATION OF MARKOV CHAINS

The term diffusion approximation is used in at least two different ways. One is in the context of problems described in Section 7, i.e., diffusion that emerges as a result of more and more rapidly varying fluctuations. The other is the passage from a motion by discontinuous movements, a random walk, to a continuous motion, a diffusion. In the latter problems ergodic theory does not enter. There are actually two ways in which diffusion approximations of the second kind arise. One is the random walk (or Markov chain) to diffusion limit when the random walk is given at first. In the other the diffusion is given, a consistent difference approximation is defined, and then one studies the convergence of the difference approximation. The literature on these problems is huge. We mention here the work of Khinchine [42],

Skorohod [43], Stroock and Varadhan [6] using martingales, John [44], and of Kushner [45].

We now describe briefly these problems for comparison purposes with the "ergodic" diffusion theory.

Let $P_\epsilon(x, dy)$ be the transition probability of a discrete time Markov process $x^\epsilon(n)$, $n = 0, 1, 2, \ldots$ on R^n with $\epsilon > 0$ a parameter. We are interested in the convergence of $x^\epsilon([t/\epsilon])$, where $[\ ] =$ integer part of, as $\epsilon \to 0$ to a diffusion process $x(t)$. For this purpose we assume the following.

For any $\delta > 0$,

$$\lim_{\epsilon \downarrow 0} \sup_x \frac{1}{\epsilon} \int_{|x-y|>\delta} P_\epsilon(x, dy) = 0 \qquad \text{(Lindeberg condition)}, \quad (8.1)$$

there is a smooth and bounded n-vector function $b(x)$ such that

$$\lim_{\epsilon \downarrow 0} \sup_x \left| \frac{1}{\epsilon} \int_{|y-x| \leq \delta} P_\epsilon(x, dy)(y - x) - b(x) \right| = 0, \qquad (8.2)$$

there is a smooth, bounded, nonnegative definite $n \times n$ matrix function $a(x)$ such that*

$$\lim_{\epsilon \downarrow 0} \sup_x \left| \frac{1}{\epsilon} \int_{|y-x| \leq \delta} P_\epsilon(x, dy)(y - x) \otimes (y - x) - a(x) \right| = 0. \quad (8.3)$$

Let T_ϵ be the transition operator corresponding to P_ϵ and suppose that it maps $\hat{C}(R^n) \to \hat{C}(R^n)$ ($\hat{C}$ = Banach space of bounded continuous functions that vanish at infinity). Let $T(t)$ be the semigroup of diffusion process with diffusion matrix $a(x)$ and drift vector $b(x)$. From results in differential equations we know that $T(t)$ exists if the coefficients are $a(x)$, $b(x)$ are smooth and $T(t) : \hat{C} \to \hat{C}$ continuously in t.

The statement of the approximation result is now that $T_\epsilon^{[t/\epsilon]} f \to T(t) f$ for each $f \in \hat{C}$, uniformly in $0 \leqslant t \leqslant t_0$ as $\epsilon \to 0$. Since smooth

* $\otimes$ is the tensor product of vectors.

functions are preserved by $T(t)$ the proof is elementary and amounts to a two-term Taylor expansion in the expression (f smooth)

$$\frac{1}{\epsilon}\left(T_\epsilon - I\right) f(x) - \mathscr{L} f(x)$$

as $\epsilon \downarrow 0$. Here $\mathscr{L}$ is the diffusion operator with coefficients $a(x)$ and $b(x)$.

The "nonsmooth" case requires compactness arguments with best results given in [6].

9. STABILITY, INVARIANT MEASURES AND LARGE DEVIATIONS FOR STOCHASTIC EQUATIONS

Consider again problem (4.9) and its first approximation (4.10) under hypothesis (4.11) so that for any $T < \infty$

$$\lim_{\epsilon \downarrow 0} \sup_{0 \leqslant t \leqslant T} |x^\epsilon(t) - \bar{x}(t)| = 0,$$

for all starting points $x \in R^n$. Suppose that

$$\bar{F}(0) = 0, \qquad F(0, t) = 0, \qquad t \geqslant 0 \tag{9.1}$$

and that the matrix $\partial \bar{F}(0)/\partial x$ has eigenvalues with negative real parts. $F(x, t)$ and $\bar{F}(x)$ are smooth vector functions of x here. Hypothesis (9.1) implies that $x = 0$ is an equilibrium point of the limit problem (4.10); in fact an asymptotically stable equilibrium point, since the variational matrix has eigenvalues with negative real parts. The question is now whether the solution $x^\epsilon(t)$, with ϵ fixed but sufficiently small, converges as $t \uparrow \infty$ to the origin if the initial point is near enough to the origin.

The answer to this question is yes; it is a result of Bogoliubov which can be found in [17] (cf. also [18]).

Before returning to this problem we mention some other related, more difficult questions. If the limit problem (4.10) has an orbitally stable periodic solution, does (4.9) also have a periodic solution for ϵ sufficiently small? If for $\epsilon > 0$ there are no periodic solutions, are there solutions that remain on some specific (invariant) manifold? The latter is the problem of bifurcation of a periodic solution of

(4.10) to an invariant manifold. We refer to [19] for consideration of these questions and further references to the sizable literature.

Let us consider now Bogoliubov's problem. Since

$$\frac{d\bar{x}(t)}{dt} = \bar{F}(\bar{x}(t)), \qquad \bar{x}(0) = x, \tag{9.2}$$

has $x = 0$ as a stable solution, there is a smooth function $V(x)$ in R^n that is positive definite

$$V(x) \geqslant 0, \qquad V(x) = 0 \Rightarrow x = 0 \tag{9.3}$$

and

$$\bar{F}(x)\cdot\frac{\partial V(x)}{\partial x} = \sum_{i=1}^{n} \bar{F}_i(x)\frac{\partial V(x)}{\partial x_i} \leqslant -\gamma V(x), \qquad \gamma > 0, \tag{9.4}$$

for x is a fixed neighborhood $\mathscr{D}$ of zero having compact closure.

Let $V_1(x, \tau)$ be defined by

$$V_1(x, \tau) = -\int_0^\tau [F(x, s) - \bar{F}(x)]\cdot\frac{\partial V(x)}{dx}\, ds, \tag{9.5}$$

and put

$$V^\epsilon(x, t) = V(x) + \epsilon V_1(x, t/\epsilon). \tag{9.6}$$

Since V_1 is bounded in τ (uniformly), if ϵ is sufficiently small and x is in a fixed neighborhood of zero with compact closure there are constants positive c_1 and c_2 such that for $\epsilon \leqslant \epsilon_0$,

$$c_1 V(x) \leqslant V^\epsilon(x, t) \leqslant c_2 V(x), \qquad x \in \bar{\mathscr{D}}, \tag{9.7}$$

Let $\tilde{\gamma}$ be a constant to be chosen below and consider

$$\begin{aligned}
&\left(\frac{\partial}{\partial t} + F\left(x^\epsilon(t), \frac{t}{\epsilon}\right)\cdot\frac{\partial}{\partial x} + \tilde{\gamma}\right) V^\epsilon(x, t) \\
&\quad = \tilde{\gamma} V^\epsilon(x, t) + \bar{F}(x)\cdot\frac{\partial V(x)}{\partial x} + \epsilon F\left(x, \frac{t}{\epsilon}\right)\cdot\frac{\partial V_1(x, t/\epsilon)}{\partial x} \\
&\quad \leqslant (\tilde{\gamma} c_2 - \gamma + c_3\epsilon) V(x)
\end{aligned}$$

where c_3 is a constant. If ϵ is sufficiently small there is a $\tilde{\gamma} > 0$ such that

$$\left(\frac{\partial}{\partial t} + F\left(x^\epsilon(t), \frac{t}{\epsilon}\right) \cdot \frac{\partial}{\partial x} + \tilde{\gamma}\right) V^\epsilon(x, t) \leqslant 0, \qquad x \in \bar{\mathscr{D}}. \tag{9.8}$$

Thus,

$$e^{\tilde{\gamma}t} V(x^\epsilon(t)) \leqslant e^{\tilde{\gamma}t} \frac{1}{c_1} V^\epsilon(x^\epsilon(t), t) \leqslant \frac{c_2}{c_1} V(x), \qquad x \in \bar{\mathscr{D}}, \qquad t \geqslant 0, \tag{9.9}$$

from which Bogoliubov's result follows since $V(x)$ is positive definite.

The above result can be generalized immediately to the stochastic problem (4.16) or (7.1).

Consider (4.16) under the hypothesis that the recurrent potential kernel (4.20) is well defined and with $\bar{F}(x)$ of (4.17) satisfying

$$\bar{F}(0) = 0, \qquad F(0, y) = 0. \tag{9.10}$$

Suppose there is a positive definite Lyapounov function $V(x)$ satisfying (9.3) and (9.4). Then there is an $\epsilon_0 > 0$ such that for given $\eta_1 > 0$ and $\eta_2 > 0$

$$P_{x,y}\{|x^\epsilon(t)| \leqslant \eta_2 e^{-\tilde{\gamma}t}, t \geqslant 0\} \geqslant 1 - \eta_1, \qquad \epsilon < \epsilon_0,$$

provided $|x(0)| = |x| < \delta$ and where $\tilde{\gamma} > 0$ is a constant and $y \in S$.

The proof of this is almost identical to the one just given taking into account the modifications necessary to estimate the extra martingale term (as in (4.22)) using Kolmogorov's inequality.

Lyapounov methods, like the above, for Itô stochastic equations are discussed in [3], [46] and [47], among other places. Extension of the method to stochastic Bogoliubov-like problems are given in [48]. In [48] only (7.1) is treated, it being the more difficult of the two ((4.16), (7.1)) situations.

In the stochastic context, results concerning stability and bifurcation of periodic solutions have not been analyzed as far as we know.

Let us now consider (6.2) and its limit equation (6.6). Suppose* that $\bar{F}(0) = 0$ again and take the solution $\bar{x}(t) \equiv 0$ in (6.6) so that the fluctuation process is time homogeneous. Since the forcing term

$$\int_0^t \sigma(0)\, dw(s)$$

is present, the process $z(t)$ cannot tend to a fixed limit as $t \to \infty$. At best, we may expect that as $t \to \infty$ it reaches a stationary distribution. We can easily assess the situation since $z(t)$ is Gaussian and time homogeneous. In fact if $Q = \partial \bar{F}(0)/\partial x$ has eigenvalues with negative real parts, $z(t)$ has a unique stationary distribution which is Gaussian with mean zero and covariance (cf. (6.9))

$$\int_0^\infty e^{Qt} a(0) e^{Q^T t}\, dt.$$

The problem is now to find out if for ϵ sufficiently small but fixed, $z^\epsilon(t)$, the solution of (6.2) (with $\bar{x}(t) \equiv 0$) and $F(0, y) \equiv 0$) also has a stationary or invariant distribution as $t \to \infty$ (which will not be Gaussian, however).

From results in [48] one can conclude that this is sometimes true, but the stationary distribution when ϵ is positive may not be unique. Any sequence of such invariant distributions tends to the Gaussian limit as $\epsilon \to 0$, however.

In many important problems $z^\epsilon(t)$ of (6.2) fails to behave as $t \to \infty$, $\epsilon > 0$, like the limit as $t \to \infty$ of (6.6), no matter how small $\epsilon > 0$ is chosen. Stated another way, the limits $t \to \infty$ and $\epsilon \to 0$ cannot be interchanged without changing the result. The reasons for this are clear from the nature of the approximation (6.6) which is local—in the neighborhood of a fixed, stable limit orbit. When $t \to \infty$ with $\epsilon > 0$ the global motion governed by the vector field F enters the picture and large deviations (which are not negligible when $t \to \infty$) may distort the local picture entirely.

In Section 12 we return to the question of large deviations in a different context.

---

* The $\omega \in \Omega$, $(\Omega, \mathscr{F}, P)$ a probability space, is omitted.

10. TRANSPORT PROBLEMS

Transport problems are of the form (7.1) with $y(t)$ a jump Markov process so that its infinitesimal generator A on $B(S)$ has the form

$$Af(y) = q(y)\int_S \pi(y, dz)f(z) - q(y)f(y). \tag{10.1}$$

Here we assume that

$$0 < q_l \leqslant q(y) \leqslant q_u < \infty \tag{10.2}$$

and that the transition probability function $\pi(y, B)$ satisfies Doeblin's condition relative to some reference measure on S. Thus the jump process $y(t)$ on S governed by A is strongly ergodic. The function $q(y)$ is the local rate with which jumps take place and $\pi(y, B)$ is the jump transition probability at the instant that a jump occurs.

The process $(x^\epsilon(t), y^\epsilon(t))$ governed by (7.1) is a Markov process on $R^n \times S$ with generator

$$\mathscr{L}^\epsilon = \frac{1}{\epsilon^2} A + \frac{1}{\epsilon} F(x, y)\cdot\frac{\partial}{\partial x}. \tag{10.3}$$

From the point of view of differential equations we are dealing with the integrodifferential Kolmogorov equation

$$\begin{aligned}\frac{\partial u^\epsilon(t, x, y)}{\partial t} &= \frac{1}{\epsilon} F(x, y)\cdot\frac{\partial u^\epsilon(t, x, y)}{\partial x} \\ &\quad + \frac{q(y)}{\epsilon^2}\int_S \pi(y, dz)u^\epsilon(t, x, z) - \frac{q(y)}{\epsilon^2} u^\epsilon(t, x, z), \qquad (10.4)\\ &\qquad t > 0, \qquad x \in R^n, \qquad y \in S,\\ u^\epsilon(0, x, y) &= f(x, y).\end{aligned}$$

This is a transport equation, a basic equation in many areas of physics (cf. [52] and the references cited there).

In the physical contexts one deals with particle densities, hence forward equations, but in any case the underlying probability measure is what one wants to analyze. In physical contexts, S is usually R^n or a subset thereof and y is thought of as velocity of a moving particle. One also takes, usually, $F(x, y) = y$ if the medium without the scattering is not refracting. The parameter ϵ in (10.4) corresponds to the mean free time (or mean free path) between collisions or discontinuous changes in the velocity. Equation (10.4) subject to (7.2) is scaled for passage to the diffusion limit.

Equation (10.4) is also called a conservative transport equation because $A1 = 0$. In many interesting problems the operator A generates a semigroup $T(t)$ which is positivity preserving (like a Markov semigroup) but $T(t)1 \neq 1$ so it is not a Markov semigroup. Such problems are called nonconservative and the generator A can be written a sum of a Markov generator $\tilde{A}$ plus a potential term V (multiplication operation by V). The sign of V corresponds to creation or destruction (locally) of particles.

Depending on whether or not the term V is large (of $O(1)$) or small, the corresponding asymptotic analysis of (10.4) changes. If $V = O(\epsilon^2)$ there are no essential changes in (10.4) as it stands. If $V = O(1)$ then in general there will be no limit as $\epsilon \to 0$ in the usual sense. Let $T_V(t)$ be the semigroup generated by $\tilde{A} + V$ on S, suppose that

$$\lim_{t \uparrow \infty} \frac{1}{t} \log \|T_V(t)\| = 0^*$$

and that it satisfies a Doeblin condition (cf. (3.11)). Then [51] there is a positive function $\phi(y)$ such that $(\tilde{A} + V)\phi(y) = 0$. If we now consider not $u^\epsilon(t, x, y)$ but $v^\epsilon(t, x, y) \equiv u^\epsilon(t, x, y)/\phi(y)$ then it satisfies a conservative transport equation and we are back to (10.4).

Note that in (10.4) the initial data depends on both x and y while in (7.6) we did not allow this. Since in the diffusion limit dependence on y disappears, it is clear that an initial layer develops near $t = 0$.

Let $\bar{\mathscr{L}}$ denote the operator defined by (7.17) with (a_{ij}) and (b_j) as in (7.15) and (7.16) and with A now as in (10.1). Let

$$\bar{f}(x) = \int \bar{P}(dy) f(x, y), \tag{10.5}$$

$$u^{IL}(\tau, y; x) = \int P(\tau, y, dz)[f(z, x) - \bar{f}(x)]. \tag{10.6}$$

---

* This limit exists by subadditivity of $\log \|T_V(t)\|$.

Then, it is easily seen that

$$\left| u^{\epsilon}(t, x, y) - u(t, x) - u^{IL}\left(\frac{t}{\epsilon^2}, y; x\right) \right| \to 0 \tag{10.7}$$

uniformly in $(x, y) \in R^n \times S$ and $0 \leqslant t \leqslant T < \infty$ as $\epsilon \to 0$. Here $u(t, x)$ satisfies the diffusion equation

$$\frac{\partial u(t, x)}{\partial t} = \bar{\mathscr{L}} u(t, x), \qquad t > 0$$

$$u(0, x) = \bar{f}(x). \tag{10.8}$$

Note that $u^{IL}(\tau, y; x)$ satisfies the initial layer equation

$$\frac{\partial u^{IL}}{\partial \tau}(\tau, y; x) = A u^{IL}(\tau, y; x), \qquad \tau > 0$$

$$u^{IL}(0, y; x) = f(x, y) - \bar{f}(x). \tag{10.9}$$

The ergodic properties of $y(t)$ generated by A imply that the third term in (10.7), the initial layer correction, is negligible away from an $O(\epsilon^2)$ neighborhood of $t = 0$.

Nonuniform behavior near spatial boundaries, or boundary layers, is discussed in the next section.

11. APPROXIMATIONS IN BOUNDED DOMAINS, BOUNDARY CONDITIONS AND BOUNDARY LAYERS

Let $\mathscr{D} \subset R^n$ be a bounded open set with smooth boundary $\partial\mathscr{D}$. For each $x \in \partial\mathscr{D}$ let S^+ and S^- be a decomposition of the state space S as follows

$$S^{\pm} = \{y \in S : \hat{n}(x) \cdot F(x, y) \gtreqless 0, x \in \mathscr{D}\}. \tag{11.1}$$

Here $\hat{n}(x)$ denotes the unit outward normal to $\partial\mathscr{D}$ at x.

We consider now the boundary value problem

$$\mathscr{L}^{\epsilon} u^{\epsilon}(x) = \frac{1}{\epsilon} F(x, y) \cdot \frac{\partial u^{\epsilon}(x, y)}{x} + \frac{q(y)}{\epsilon^2} \int_S \pi(y, dz) u^{\epsilon}(x, z)$$

$$- \frac{q(y) u^{\epsilon}(x, y)}{\epsilon^2} = 0,$$

$$x \in \mathscr{D}, \qquad y \in S \qquad \text{and} \qquad x \in \partial\mathscr{D}, \qquad y \in S^-, \tag{11.2}$$

$$u^{\epsilon}(x, y) = f(x, y), \qquad x \in \partial\mathscr{D}, \qquad y \in S^+. \tag{11.3}$$

This is a time independent problem and we are interested in the asymptotic behavior of the solution $u^\epsilon(x, y)$ as $\epsilon \to 0$.

First we discuss the probabilistic representation of the solution of (11.2) and its well-posedness.

Let $(x^\epsilon(t), y^\epsilon(t))$ be the Markov process on $R^n \times S$ generated by (10.3) under the usual ergodicity hypotheses on A of (10.1). Let $\bar{\mathscr{L}}$ be defined by (7.17) (via (7.15)–(7.16) with A given by (10.1)) and assume that its coefficients are smooth with at least one diagonal entry of (a_{ij}) strictly positive in $\bar{\mathscr{D}}$. We assume (7.2) holds.

Let τ be the first exit time of $x^\epsilon(t)$ from $\mathscr{D}$. From our assumptions it follows that for all ϵ sufficiently small $E_{x,y}\{\tau\} \leqslant c < \infty$. Now on the basis of the remarks in Section 2 regarding the use of the optional stopping theorem for probabilistic representations, we conclude that if

$$u^\epsilon(x, y) = E_{x,y}\{f(x^\epsilon(\tau), y^\epsilon(\tau))\}, \tag{11.4}$$

then this is well defined and is the unique solution of (11.2) (its integral equation version) with f bounded.

If f was a function of x only then, it is clear that $u^\epsilon(x, y) \to u(x)$ uniformly in $(x, y) \in \bar{\mathscr{D}} \times S$ as $\epsilon \to 0$ where

$$\begin{aligned} \bar{\mathscr{L}}u(x) &= 0, \qquad x \in \mathscr{D} \\ u(x) &= f(x), \qquad x \in \partial\mathscr{D} \end{aligned} \tag{11.5}$$

is an elliptic boundary value problem. The proof of this is a straightforward extension of the one in Section 7 given that $E\{\tau\} \leqslant c < \infty$. The latter is proved by an argument like the one in Section 9.

How does $u^\epsilon(x, y)$ behave in the general case? We must consider more closely what happens near the boundary $\partial\mathscr{D}$.

We define stretched boundary layer coordinates near $\partial\mathscr{D}$ so that $x \to (\gamma, \eta)$ where (for $x \in \mathscr{D}$)

$$x = \xi(\gamma) + \epsilon\eta\hat{n}(\gamma), \qquad \eta \leqslant 0.$$

Here $\xi(\gamma)$ is the local parametric representation of the surface $\partial\mathscr{D}$, $\gamma = (\gamma_1, \gamma_2, \ldots, \gamma_{n-1}) \in R^{n-1}$, and $\eta \leqslant 0$ is distance from the boundary. With each x near enough to $\partial\mathscr{D}$, $\xi(\gamma)$ $(=\xi(\gamma(x))$ since $\gamma = \gamma(x))$ is the unique point on $\partial\mathscr{D}$ with $x - \xi(\gamma)$ parallel to the normal at the same point. In local boundary layer coordinates,

$$\frac{\partial}{\partial x_i} \to \frac{1}{\epsilon}\hat{n}_i(\gamma)\frac{\partial}{\partial \eta} + \sum_{k=1}^{n-1} \Gamma_{ik}^{\epsilon}(\gamma)\frac{\partial}{\partial \gamma_k}, \qquad i = 1, 2, \ldots, n,$$

where $\hat{n}_i(\gamma)$ (or $\hat{n}_i(x)$, $x \in \partial\mathscr{D}$) are the components of the normal vector and $\Gamma_{ik}^{\epsilon}(\gamma)$ are functions that depend on the geometry of $\partial\mathscr{D}$ (its principal curvatures) and are regular in $\epsilon \geqslant 0$.

Next we express the generator $\mathscr{L}^{\epsilon}$ near $\partial\mathscr{D}$ in terms of (γ, η). It has the form $(1/\epsilon^2)\mathscr{L}_{BL}$ + less singular in ϵ terms where $(x \in \partial\mathscr{D})$

$$\mathscr{L}_{BL} = \hat{n}(x)\cdot F(x, y)\cdot\frac{\partial}{\partial \eta} + q(y)\int \pi(y, dz) - q(y). \tag{11.6}$$

It is easy to see that $\mathscr{L}_{BL}$ generates a Markov process $(H(t), y(t))$ on $(-\infty, \infty) \times S$. In fact, since $x \in \partial\mathscr{D}$ is merely a parameter in $\mathscr{L}_{BL}$, if we let

$$v(y; x) = \hat{n}(x)\cdot F(x, y), \tag{11.7}$$

then $y(t)$ is the usual process on S and

$$H(t) = \eta + \int_0^t v(y(s); x)\, ds, \tag{11.8}$$

with η the starting point.

Let $\bar{\tau}$ be the first exit time of $H(t)$ from $(-\infty, 0]$ given that $H(0) = \eta < 0$ or $\eta = 0$ and $y \in S^-$ $(v(y) \leqslant 0$ on $S^-)$. Hypothesis (7.2) implies that $\bar{\tau}$ is a proper random variable and hence we may consider the exit random variable $y(\bar{\tau})$ which takes values in S^+. Let $P_{BL}(\eta, y, B)$, $y \in S^+$, $B \subset S^+$, $\eta \geqslant 0$ be defined by

$$P_{BL}(\eta, y, B) = P\{y(\bar{\tau}) \in B \mid H(0) = -\eta, y(0) = y\}. \tag{11.9}$$

What we have just pointed out and the fact that the operator $\mathscr{L}_{BL}$ has η-independent coefficients, implies that P_{BL} is a Markov transition probability function on S^+ with η (=distance from the origin) playing the role of the time variable. The boundary layer analysis is intimately related to the ergodic properties of P_{BL} as $\eta \to \infty$.

Suppose in fact that P_{BL} is ergodic and let $\bar{P}_{BL}$ be its invariant measure (it is not necessary to assume existence of the recurrent potential unless we want exponential decay of the boundary layer corrections away from $\partial\mathscr{D}$). $\bar{P}_{BL}$ has the following meaning: it is the equilibrium (or stationary) distribution of the "velocity" variable $y(t)$ at the instant $x(t)$ touches $\partial\mathscr{D}$ while the process begins further and further in the interior.

Define

$$\bar{f}(x) = \int_{S+} \bar{P}_{BL}(dy) f(x, y), \tag{11.10}$$

and, for $\eta \leqslant 0$,

$$u^{BL}(\eta, y; x) = \int_{S+} P_{BL}(-\eta, y, dz)[f(x, z) - \bar{f}(x)], \qquad x \in \partial\mathscr{D}. \tag{11.11}$$

We may let $y \in S$ in (11.11) by extending the above definition. If the recurrent potential of P_{BL} exists then (11.11) tends to zero exponentially fast as $\eta \to -\infty$, uniformly in y (and in the parameter $x \in \partial\mathscr{D}$).

With this information at hand we return to (11.2) and (11.3). We have that

$$\limsup_{\epsilon \downarrow 0} \sup_{\substack{x \in \bar{\mathscr{D}} \\ y \in S}} \left| u^{\epsilon}(x, y) - u(x) - \zeta(x) u^{BL}\left(\frac{(x - \xi(x)) \cdot \hat{n}(x)}{\epsilon}, y; \xi(x)\right) \right| = 0.$$

Here $\zeta(x) = 1$ in a sufficiently small neighborhood of $\partial\mathscr{D}$ and zero outside a larger neighborhood and it is smooth. The function $u(x)$ is the solution of (11.5).

The proof of the above result and many other examples (reflecting boundary conditions, etc.) can be found in [52].

12. APPROXIMATION OF PROCESSES WITH SMALL DIFFUSION

We shall consider the effect of a small amount of noise on a system, when the noise is fully developed. The problems in this section are not related to the rapid fluctuation limit. They have been analyzed by a number of authors, for example in [49], [50], [53]–[58], by a variety of methods, probabilistic and analytical.

Let $(a_{ij}(x))$ and $(b_j(x))$ be a symmetric $n \times n$ positive definite matrix function of $x \in R^n$ and an n-vector function of $x \in R^n$, both assumed smooth. Let $\mathscr{D} \subset R^n$ be a bounded region with smooth boundary. We consider first the boundary value problem ($\epsilon > 0$)

$$\mathscr{L}^\epsilon u^\epsilon(x) = \epsilon \sum_{i,j=1}^n a_{ij}(x) \frac{\partial^2 u^\epsilon(x)}{\partial x_i \partial x_j} + \sum_{j=1}^n b_j(x) \frac{\partial u^\epsilon(x)}{\partial x_j}$$
$$- c(x) u^\epsilon(x) + f(x) = 0, \qquad x \in \mathscr{D}$$
$$u^\epsilon(x) = g(x), \qquad x \in \partial\mathscr{D}. \tag{12.1}$$

Here $f(x)$, $x \in \mathscr{D}$ and $g(x)$, $x \in \partial\mathscr{D}$ are given smooth functions and we assume that

$$c(x) \geqslant c_0 > 0, \tag{12.2}$$

although this is not necessary for what follows.

It is clear that (12.1) has a unique smooth solution u^ϵ and the problem is to describe its asymptotic behavior as $\epsilon \to 0$. Let $(\sigma_{ij}(x))$ be the smooth symmetric square root of $(a_{ij}(x))$ and let $x^\epsilon(t)$ be the diffusion process on R^n solving the stochastic differential equation

$$dx^\epsilon(t) = b(x^\epsilon(t))\, dt + \sqrt{2\epsilon}\,\sigma(x^\epsilon(t))\, dw(t), \qquad x^\epsilon(0) = x. \tag{12.3}$$

Let τ^ϵ be the first exit time from $\mathscr{D}$ of $x^\epsilon(t)$ given that $x^\epsilon(0) = x \in \mathscr{D}$. Then,

$$u^\epsilon(x) = E_x\left\{\int_0^{\tau^\epsilon} \left(\exp\left\{-\int_0^s c(x^\epsilon(\gamma))\, d\gamma\right\}\right) f(x^\epsilon(s))\, ds\right\}$$
$$+ E_x\left\{\left(\exp\left\{-\int_0^{\tau^\epsilon} c(x^\epsilon(\gamma))\, d\gamma\right\}\right) g(x^\epsilon(\tau^\epsilon))\right\}. \tag{12.4}$$

Depending on the nature of the trajectories of

$$dx(t) = b(x(t))\, dt, \qquad x(0) = x, \tag{12.5}$$

the formal limit of (12.3) when ϵ, the noise intensity, is zero, different types of behavior can arise. We begin with the simplest.

Assume that $b(x) \neq 0$ for $x \in \bar{\mathscr{D}}$ and let

$$\Gamma^+ \subset \partial\mathscr{D} = \{x \in \partial\mathscr{D} \mid \exists \tau > 0 \text{ and } \tilde{x} \in \mathscr{D} \text{ such that}$$
$$x(0) = \tilde{x} \text{ and } x = x(\tau) \in \partial\mathscr{D}\}. \tag{12.6}$$

Let $\Gamma^- = \partial\mathscr{D} - \Gamma^+$. The set Γ^- consists of points that can be reached along orbits of (12.5) when the time is running backward (negative) and of points that are on orbits themselves. Note also that on Γ^+, $b(x)\cdot\hat{n}(x) > 0$, where $\hat{n}(x)$ is the unit outward normal, and on Γ^-, $b(x)\cdot\hat{n}(x) \leqslant 0$. To avoid complications we assume that $b(x)\cdot\hat{n}(x) = 0$ only at two isolated points of the boundary.

It is clear that at $x \in \mathscr{D}$, $u^\epsilon(x) \to u_0(x)$ where

$$u_0(x) = \int_0^\tau \left(\exp\left\{-\int_0^s c(x(\gamma))\, d\gamma\right\}\right) f(x(s))\, ds + \left(\exp\left\{-\int_0^\tau c(x(\gamma))\, d\gamma\right\}\right) g(x(\tau)). \tag{12.7}$$

Here τ is the exit time of the deterministic orbits (12.5). Evidently $u_0(x) \neq g(x)$, $x \in \Gamma^-$ (except the two points) and so $u_0(x)$ does not satisfy the boundary condition in the formal limit $\epsilon = 0$ in (12.1). We must introduce a boundary layer correction.

We employ the notation of Section 11. Near $\partial\mathscr{D}$ we represent x by (γ, η) coordinates where

$$x = \xi(\gamma) + \epsilon\eta\hat{n}(\gamma), \qquad \eta \leqslant 0 \tag{12.8}$$

(cf. (11.4), (11.5)). In these coordinates the operator $\mathscr{L}^\epsilon$ on the left side of (12.1) takes the form

$$\frac{1}{\epsilon}\left[\sum_{i,j=1}^{n} a_{ij}(\xi(\gamma))\hat{n}_i(\gamma)\hat{n}_j(\gamma)\frac{\partial^2}{\partial\eta^2} + \sum_{j=1}^{n} b_j(\xi(\gamma))\hat{n}_j(\gamma)\frac{\partial}{\partial\eta}\right] + O(1), \tag{12.9}$$

where $O(1)$ stands for a differential operator in $\partial/\partial\eta$, $\partial/\partial\gamma_k$ with coefficients that are regular in $\epsilon \geqslant 0$. Let

$$\tilde{a}(x) = \sum_{i,j=1}^{n} a_{ij}(x)\hat{n}_i(x)\hat{n}_j(x), \qquad x \in \partial\mathscr{D}$$

$$\tilde{b}(x) = \sum_{j=1}^{n} b_j(x)\hat{n}_j(x), \qquad x \in \partial\mathscr{D} \tag{12.10}$$

and

$$\mathscr{L}_{BL} = \tilde{a}(x)\frac{\partial^2}{\partial\eta^2} + \tilde{b}(x)\frac{\partial}{\partial\eta}. \tag{12.11}$$

Let $u^{BL}(\eta; x)$ be the solution of

$$\begin{aligned} \mathscr{L}_{BL}u^{BL}(\eta; x) &= 0, \qquad \eta < 0 \\ u^{BL}(0; x) &= g(x), \\ u^{BL}(-\infty, x) &= 0, \qquad x \in \Gamma^-. \end{aligned} \tag{12.12}$$

Clearly

$$u^{BL}(\eta; x) = g(x)e^{-\tilde{b}(x)\eta/\tilde{a}(x)}, \qquad x \in \Gamma^- \tag{12.13}$$

where, we recall, $\tilde{b}(x) < 0$ on Γ^-.

Let $\zeta(x)$ be equal to one in a neighborhood θ_1 of Γ^-, equal to zero outside of $\theta_2 \supset \theta_1$ and smooth. Let θ_2 be small enough so that the coordinates (12.8) are valid. Then,

$$\limsup_{\epsilon \downarrow 0} \sup_{x \in \mathscr{D}} \left| u^\epsilon(x) - u_0(x) - \zeta(x)u^{BL}\left(\frac{(x - \xi(x))\cdot\hat{n}(x)}{\epsilon}; x\right) \right| = 0. \tag{12.14}$$

To prove this result we let

$$w^\epsilon = u^\epsilon - u_0 - \zeta u^{BL}. \tag{12.15}$$

Then $w^\epsilon(x) = 0$, $x \in \partial\mathscr{D}$ by construction. Moreover, with $\mathscr{L}^\epsilon$ the operator on the left side of (12.1), we have

$$\begin{aligned} \mathscr{L}^\epsilon w^\epsilon &= -f - \mathscr{L}^\epsilon u_0 - \mathscr{L}^\epsilon(\zeta u^{BL}) \\ &= -\epsilon \sum_{i,j=1}^{n} a_{ij} \frac{\partial^2 u_0}{\partial x_i \partial x_j} - \zeta\left(\mathscr{L}^\epsilon - \frac{1}{\epsilon}\mathscr{L}^{BL}\right)u^{BL} \\ &\quad - [\mathscr{L}^\epsilon(\zeta u^{BL}) - \zeta\mathscr{L}^\epsilon u^{BL}). \end{aligned} \tag{12.16}$$

By the decay of u^{BL} to zero as $\eta \to -\infty$ it follows that the right-hand side of (12.16) is uniformly bounded and at each $x \in \mathscr{D}$ it tends to zero as $\epsilon \to 0$. In view of (12.2) and the maximum principle the result (12.14) follows (actually one must look in a bit more detail at what happens near the two special points of Γ^- where $\hat{n}\cdot b = 0$; continuity, however, prevents difficulties).

As another example we consider a Neumann problem as follows

$$\begin{aligned} \mathscr{L}^\epsilon u^\epsilon(x) + f(x) &= 0, \qquad x \in \mathscr{D}, \\ \gamma(x)\cdot\nabla u^\epsilon(x) + \alpha(x)u^\epsilon(x) - g(x) &= 0, \qquad x \in \partial\mathscr{D}. \end{aligned} \tag{12.17}$$

Here $\gamma(x)$ is a vector field defined on $\partial\mathscr{D}$, smooth, so that

$$\gamma(x)\cdot\hat{n}(x) \leqslant \gamma_0 < 0, \qquad x \in \partial\mathscr{D}, \tag{12.18}$$

i.e., $\gamma(x)$ points inward and is uniformly nontangential. In addition we assume that

$$\alpha(x) \geqslant \alpha_0 > 0. \tag{12.19}$$

The stochastic process associated with (12.17) is a reflected diffusion process. In terms of stochastic differential equations it can be defined as the solution of

$$dx^\epsilon(t) = b(x^\epsilon(t))\,dt + \sqrt{2\epsilon}\sigma(x^\epsilon(t))\,dw(t) + \gamma(x^\epsilon(t))\,dN^\epsilon(t), \tag{12.20}$$
$$x^\epsilon(0) = x \in \mathscr{D}.$$

Here $N^\epsilon(t)$ is a continuous, nondecreasing, nonanticipating functional of Brownian motion such that $N^\epsilon(t)$ increases only when $x^\epsilon(t) \in \partial\mathscr{D}$. In terms of martingales [8], the approach we follow, the measure P_x^ϵ on $C([0,\infty); \bar{\mathscr{D}})$ associated with $\{\mathscr{L}^\epsilon, \gamma\}$ of (12.17) is the solution of the submartingale problem

(i) $P_x^\epsilon\{x(0) = x\}$,

(ii) $h(x(t)) - \int_0^t \mathscr{L}^\epsilon h(x(s))\,ds$ is a submartingale relative to P_x^ϵ for

each $h \in C^2(\bar{\mathscr{D}})$ with $\gamma(x)\cdot\nabla h(x) \geqslant 0$. (12.21)

For each $h(x) \in C^2(\bar{\mathscr{D}})$ there is an increasing process $N_h^\epsilon(t)$ so that

$$h(x(t)) - \int_0^t \mathscr{L}^\epsilon h(x(s))\,ds - N_h^\epsilon(t)$$

is a martingale relative to P_x^ϵ (when the measure is explicitly displayed we do not use superscript ϵ on $x(\cdot)$). To normalize this increasing process we let $\phi(x)$ be a support function for the smooth domain $\mathscr{D}$, i.e.,

$$\mathscr{D} = \{x \mid \phi(x) > 0\}$$
$$\partial\mathscr{D} = \{x \mid \phi(x) = 0\} \tag{12.22}$$

and $|\nabla\phi(x)| \geqslant 1$, $x \in \partial\mathscr{D}$. Let $N^\epsilon(t)$ be the increasing process corresponding to $h = \phi$ above. For any other smooth h we have that

$$N_h^\epsilon(t) = \int_0^t \frac{\gamma(x(s))\cdot\nabla h(x(s))}{\gamma(x(s))\cdot\nabla\phi(x(s))}\, dN^\epsilon(s), \tag{12.23}$$

which is well defined by (12.18). Let us, in particular, normalize the choice of ϕ so that $\gamma\cdot\nabla\phi = 1$.

Now the solution of (12.17) can be expressed as follows.

$$\begin{aligned} u^\epsilon(x) = {} & E_x\left\{\int_0^\infty \left(\exp\left\{-\int_0^t c(x^\epsilon(s))\,ds - \int_0^t \alpha(x^\epsilon(s))\,dN^\epsilon(s)\right\}\right) f(x^\epsilon(t))\,dt\right\} \\ & + E_x\left\{\int_0^\infty \left(\exp\left\{-\int_0^t c(x^\epsilon(s))\,ds - \int_0^t \alpha(x^\epsilon(s))\,dN^\epsilon(s)\right\}\right)\right. \\ & \left. \times\, g(x^\epsilon(s))\,dN^\epsilon(s)\right\}. \end{aligned} \tag{12.24}$$

That this is a well-defined expression follows easily from the boundedness of f and g and from (12.19), (12.2).

Let us return to the perturbation expansion. We decompose $\partial\mathscr{D}$ into Γ^+ and Γ^- as before. The boundary layer near Γ^- is essentially the same as before. What is interesting now is the interior limit solution $u_0(x)$, valid on $\mathscr{D} \cup \Gamma^+$ and satisfying

$$\begin{aligned} b(x)\cdot\nabla u_0(x) - c(x)u_0(x) + f(x) &= 0, \qquad x \in \mathscr{D} \\ \gamma(x)\cdot\nabla u_0(x) - \alpha(x)u_0(x) + g(x) &= 0, \qquad x \in \partial\mathscr{D}. \end{aligned} \tag{12.25}$$

Any vector on $\partial\mathscr{D}$ can be decomposed into the sum of a vector along the normal $\hat{n}$ and a vector tangent to $\partial\mathscr{D}$. For $\gamma(x)$ and $b(x)$ we write

$$\gamma(x) = \gamma(x)\cdot\hat{n}(x)\hat{n}(x) + \gamma_t(x), \qquad x \in \partial\mathscr{D} \tag{12.26}$$

$$b(x) = b(x)\cdot\hat{n}(x)\hat{n}(x) + b_t(x), \qquad x \in \partial\mathscr{D} \tag{12.27}$$

where $\gamma_t(x)$ and $b_t(x)$ are defined by (12.26) and (12.27). Using these formulas in (12.75) we obtain

$$\begin{aligned} \gamma(x)\cdot\hat{n}(x)\frac{\partial u_0(x)}{\partial\hat{n}(x)} + \gamma_t(x)\cdot\nabla u_0(x) - \alpha(x)u_0 + g(x) &= 0, \qquad x \in \Gamma^+, \\ b(x)\cdot\hat{n}(x)\frac{\partial u_0(x)}{\partial\hat{n}(x)} + b_t(x)\cdot\nabla u_0(x) - c(x)u_0(x) + f(x) &= 0, \qquad x \in \Gamma^+. \end{aligned} \tag{12.28}$$

Since $b \cdot \hat{n} > 0$ and $\gamma \cdot \hat{n} < 0$ on $x \in \Gamma^+$ we may eliminate $\partial u_0(x)/\partial \hat{n}(x)$ and obtain an equation for a motion entirely confined to Γ^+

$$\left(\frac{b_t(x)}{b(x)\cdot\hat{n}(x)} - \frac{\gamma_t(x)}{\gamma(x)\cdot\hat{n}(x)}\right)\cdot\nabla u_0(x) - \left(\frac{c(x)}{b(x)\cdot\hat{n}(x)} - \frac{\alpha(x)}{\gamma(x)\cdot\hat{n}(x)}\right)u_0$$
$$+ \left(\frac{f(x)}{b(x)\cdot\hat{n}(x)} - \frac{g(x)}{\gamma(x)\cdot\hat{n}(x)}\right) = 0, \qquad x \in \Gamma^+. \quad (12.29)$$

The validity of the first approximation $u_0(x)$ satisfying the first equation in (12.25) in the interior and (12.29) on Γ^+ (+boundary layers on Γ^-) depends crucially on the nature of the tangential vector field

$$F_t(x) = \frac{b_t(x)}{b(x)\cdot\hat{n}(x)} - \frac{\gamma_t(x)}{\gamma(x)\cdot\hat{n}(x)}, \qquad x \in \Gamma^+. \quad (12.30)$$

If $F_t(x)$ has one stable equilibrium point in Γ^+, then the expansion is valid much as in the case of the Dirichlet problem.

If, however, $F_t(x)$ has unstable equilibrium points or other singular behavior, interior boundary layers will develop. We shall not pursue this further here but we shall pass to our third and last example.

Consider the problem

$$\mathscr{L}^\epsilon u^\epsilon(x) = \epsilon \sum_{i,j=1}^{n} a_{ij}(x)\frac{\partial^2 u^\epsilon(x)}{\partial x_i \partial x_j} + \sum_{j=1}^{n} b_j(x)\frac{\partial u^\epsilon(x)}{\partial x_j} + f(x) = 0, \qquad x \in \mathscr{D}$$
$$u^\epsilon(x) = g(x), \qquad x \in \partial\mathscr{D}, \quad (12.31)$$

where $\mathscr{D} \subset R^n$ is bounded and $\partial\mathscr{D}$ is smooth and f and g are given. Assume that

$$\sum_{i,j=1}^{n} a_{ij}(x)\xi_i\xi_j \geqslant \alpha_0|\xi|^2, \qquad \alpha_0 > 0, \quad (12.32)$$

and that

$$b(0) = 0, \qquad \left.\frac{\partial b(x)}{\partial x}\right|_{x=0} \equiv B \quad (12.33)$$

has eigenvalues with negative real points. Assume finally that

$$b(x)\cdot\hat{n}(x) < 0, \qquad x \in \partial\mathscr{D} \quad (12.34)$$

where $\hat{n}(x)$ is the unit outward normal. Clearly, if $x^\epsilon(t)$ is the solution of (12.3) and τ^ϵ the first exit time (a proper random variable) then,

$$u^\epsilon(x) = E_x\left\{\int_0^{\tau^\epsilon} f(x^\epsilon(s))\,ds\right\} + E_x\{g(x^\epsilon(\tau^\epsilon))\}. \qquad (12.35)$$

The problem here is different, and much more difficult, than the previous ones because all of the boundary is the set Γ^-; there is no exit possible in the limit. In the case $f \equiv 0$, $u^\epsilon(x)$ tends as $\epsilon \to 0$ to a constant independent of x (for each x in the interior) which is a functional of the data $g(x)$ on the boundary. We shall consider here a formal analysis that leads to the correct result in this case [58]. Proofs can be found in [55], in a special case, which can be adapted in general; they are much more difficult than the above.

The solution u^ϵ tends to a constant c in the interior. Near the boundary, in local boundary layer coordinates we have the boundary layer correction (cf. (12.10)–(12.13)).

$$(g(x) - c)e^{-\tilde{b}(x)\eta/\tilde{a}(x)}, \qquad x \in \partial\mathscr{D}, \qquad (\tilde{b}(x) \leqslant 0,\ \eta \leqslant 0). \qquad (12.36)$$

Thus,

$$u^\epsilon(x) \sim c + (g - c)e^{-\tilde{b}(x-\xi)\cdot\hat{n}/\epsilon\tilde{a}} \qquad (12.37)$$

is the first term of a composite expansion with the constant c still undetermined. Carrying out the expansion to higher order will not allow for the determination of c so we determine it as follows [58].

Let $\mathscr{L}^{\epsilon^*}$ be the operator adjoint to $\mathscr{L}^\epsilon$ in (12.31). By the WKB (or geometrical optics) method [58] we construct an asymptotic solution w^ϵ of

$$\mathscr{L}^{\epsilon^*} w^\epsilon = 0$$

with w^ϵ normalized to equal one at the origin. It has the form

$$w^\epsilon = e^{\phi/\epsilon}(w_0 + \cdots), \qquad (12.38)$$

where both ϕ and w_0 can be computed readily by solving the eikonal (for ϕ) equation and the transport equation for w_0. We also have

$$0 = \int_{\mathscr{D}} [w^\epsilon \mathscr{L}^\epsilon u^\epsilon - u^\epsilon \mathscr{L}^{\epsilon *} w^\epsilon]$$

$$= \int_{\delta\mathscr{D}} \left[w^\epsilon \sum_{i,j=1}^{n} a_{ij}\hat{n}_i \frac{\partial u^\epsilon}{x_j} - u^\epsilon \sum_{i,j=1}^{n} a_{ij}\hat{n}_i \frac{\partial w^\epsilon}{\partial x_j} + \sum_{j=1}^{n} b_j(x)\hat{n}_j(x) u^\epsilon w^\epsilon \right] ds. \tag{12.39}$$

Now we insert in the identity (12.39) the expansions (12.38) and (12.37), solve for c, and take the limit $\epsilon \downarrow 0$. Thus c is obtained as the limit of the ratio of two Laplace-type integrals and this solves (formally) the problem of determining the distribution of exit points in cases (12.33), (12.34).

13. NONERGODIC DRIVING PROCESSES AND LOCAL TIME

We return now to the analysis of differential equations with rapidly fluctuating components. We shall examine briefly what happens when in (4.16) the process $y(t)$ (with $y^\epsilon(t)$ scaled in a manner other than (4.17)) on S is not ergodic but null recurrent [59]. We restrict attention to the case $S = R^1$, $y(t) =$ Brownian motion and $F(x, y)$ has compact support in y for all $x \in R$.

Let

$$\bar{F}(x) = \int_{-\infty}^{\infty} F(x, y)\, dy \tag{13.1}$$

and let $\bar{x}(t)$ satisfy the equation

$$\frac{d\bar{x}(t)}{dt} = \bar{F}(\bar{x}(t)), \qquad \bar{x}(0) = x. \tag{13.2}$$

Let $l_0(t)$ be the local time of Brownian motion $y(t)$ with $y(0) = y$. That is, formally,

$$l_0(t) = \int_0^t \delta(y(s))\, ds,$$

or by Tanaka's formula [4]

$$l_0(t) = 2y^+(t) - 2\int_0^t H(y(s))\, dy(s), \tag{13.3}$$

where $y^+ = y \vee 0$, $H(y) = 1$ if $y \geqslant 0$, $H(y) = 0$, $y < 0$. The formula (13.3) may serve as definition of (13.3) while the formal expression involving the delta function explains the terminology.

Let $x^\epsilon(t)$ be the process which is the solution of (7.1) with the above assumptions and with

$$y^\epsilon(t) = y(t/\epsilon^2). \tag{13.4}$$

Then $(x^\epsilon(t), y(t))$ converses (weakly) to $(\bar{x}(l_0(t)), y(t))$ as $\epsilon \to 0$ and $0 \leqslant t \leqslant T < \infty$.

To see what this result implies we look at the case $F(x, y) = \chi_A(y)$ with $0 \in A$. Then,

$$\begin{aligned} x^\epsilon(t) &= x + \int_0^t \frac{1}{\epsilon}\, \chi_A(y^\epsilon(s))\, ds \\ &= x + \int_0^t \frac{1}{\epsilon}\, \chi_A(y(s/\epsilon^2))\, ds \\ &\sim x + \int_0^t \frac{1}{\epsilon}\, \chi_A\!\left(\frac{1}{\epsilon}\, y(s)\right) ds \qquad \text{(Brownian scaling)} \\ &= x + \int_0^t \frac{1}{\epsilon}\, \chi_{\epsilon A}(y(s))\, ds \\ &\sim x + \int_0^t \delta(y(s))\, ds \qquad \text{(formally).} \end{aligned}$$

The appearance of the local time is clear in this example. The general result has locally the same character.

In the event $\bar{F}(x) \equiv 0$ one can obtain results analogous to those of Section 7 where the limit process is a diffusion run at the local time of a suitably scaled driving process. More details can be found in [59].

14. STOCHASTIC TWO-POINT BOUNDARY VALUE PROBLEMS

The asymptotic methods that deal with initial value problems can be adapted to deal with two-point boundary value problems. We shall do this for a general linear problem, an example of which follows in the next two sections. The considerations here are deterministic and of a formal algebraic type.

Let $x(t)$ be a vector of dimension $2n$ with the first n components denoted by $x_1(t)$ and the last n components denoted by $x_2(t)$. It satisfies the two-point boundary value problem

$$\frac{d}{dt}\begin{pmatrix} x_1(t) \\ x_2(t) \end{pmatrix} = \begin{pmatrix} A_{11}(t) & A_{12}(t) \\ A_{21}(t) & A_{22}(t) \end{pmatrix}\begin{pmatrix} x_1(t) \\ x_2(t) \end{pmatrix}, \qquad t_1 < t < t_2, \tag{14.1}$$

$$B_{11}x_1(t_1) + B_{12}x_2(t_2) = b_1,$$

$$C_{11}x_1(t_2) + C_{12}x_2(t_2) = c_1. \tag{14.2}$$

Here the $2n \times 2n$ matrix $A(t)$ is given along with the $n \times n$ matrices B_{11}, B_{12}, C_{11}, C_{12} and the n vectors b_1 and c_1. We assume that there exist $n \times n$ matrices B_{21}, B_{22}, C_{21}, C_{22} such that the $2n \times 2n$ matrices

$$B = \begin{pmatrix} B_{11} & B_{12} \\ B_{21} & B_{22} \end{pmatrix}, \qquad C = \begin{pmatrix} C_{11} & C_{12} \\ C_{21} & C_{22} \end{pmatrix},$$

are nonsingular. We may therefore write the boundary conditions (14.2) in the form

$$Bx(t_1) = b, \qquad Cx(t_2) = c \tag{14.3}$$

where

$$b = \begin{pmatrix} b_1 \\ \tilde{b}_1 \end{pmatrix}, \qquad c = \begin{pmatrix} c_1 \\ \tilde{c}_1 \end{pmatrix} \tag{14.4}$$

and $\tilde{b}_1$, $\tilde{c}_1$ are unknown n-vectors.

Let $Y(t, s)$ be the fundamental solution of (14.1)

$$\frac{dY(t, s)}{dt} = A(t)Y(t, s), \qquad t \geqslant s, \qquad Y(t, t) = I, \tag{14.5}$$

where I is the $2n \times 2n$ identity. We have

$$\begin{pmatrix} x_1(t_2) \\ x_2(t_2) \end{pmatrix} = \begin{pmatrix} Y_{11}(t_2, t_1) & Y_{12}(t_2, t_1) \\ Y_{21}(t_2, t_1) & Y_{22}(t_2, t_1) \end{pmatrix}\begin{pmatrix} x_1(t_1) \\ x_2(t_1) \end{pmatrix}. \tag{14.6}$$

Using (14.3) and (14.4) in (14.6) we obtain (B is nonsingular)

$$\begin{pmatrix} c_1 \\ \tilde{c}_2 \end{pmatrix} = CY(t_2, t_1)B^{-1}\begin{pmatrix} b_1 \\ \tilde{b}_2 \end{pmatrix}. \tag{14.7}$$

If we define

$$D = CY(t_2, t_1)B^{-1} \tag{14.8}$$

and assume that D_{12}^{-1} exists then we find that

$$\begin{aligned} \tilde{b}_2 &= D_{12}^{-1}[c_1 - D_{11}b_1], \\ \tilde{c}_1 &= (D_{21} - D_{22}D_{12}^{-1}D_{11})b_1 + D_{22}D_{12}^{-1}c_1. \end{aligned} \tag{14.9}$$

Once $\tilde{b}_2$ and $\tilde{c}_2$ have been obtained we can compute the solution vector $x(t)$ at any point $t_1 < t < t_2$ using the fundamental matrices $Y(t, t_1)$ or $Y(t_2, t)$. Thus, from the point of view of stochastic problems if one has information on the fundamental solution matrix $Y(t, s)$, then the solution of two-point (or multipoint) boundary value problems follows by algebraic considerations.

15. WAVE PROPAGATION IN A ONE-DIMENSIONAL RANDOM MEDIUM

Consider the two-point boundary value problem

$$\begin{aligned} &\frac{d^2u(t)}{dt^2} + k^2(1 + y(t))u(t) = 0, \qquad 0 < t < L, \\ &u(t) = e^{ikt} + Re^{-ikt}, \qquad t < 0, \\ &u(t) = Te^{ikt}, \qquad t > L, \\ &u(t), \quad \frac{du(t)}{dt} \text{ continuous.} \end{aligned} \tag{15.1}$$

Here $y(t)$ is a given real valued stochastic process, $k > 0$ is a constant, and the solution $u(t)$ is a complex valued process. The problem is in the form (15.1) in order to demonstrate its physical meaning. Namely, $u(t)$ is the wave function at location t when an incident wave of unit amplitude and with wave number k impinges on a slab of random medium whose index of refraction is $(1 + y(t))^{1/2}$. R and

T are complex random variables and are called the reflection and transmission coefficients respectively. Note that T and R depend on the statistical properties of $y(t)$ as well as k and L, the width of the random medium.

We convert (15.1) to the form (14.1) as follows. Let

$$u(t) = e^{ikt}x_1(t) + e^{-ikt}x_2(t),$$
$$\frac{du(t)}{dt} = ik[e^{ikt}x_1(t) - e^{-ikt}x_2(t)], \tag{15.2}$$

so that

$$x_1(t) = \frac{1}{2}e^{-ikt}\left(u(t) + \frac{1}{ik}\frac{du(t)}{dt}\right),$$
$$x_2(t) = \frac{1}{2}e^{ikt}\left(u(t) - \frac{1}{ik}\frac{du(t)}{dt}\right). \tag{15.3}$$

Then, we have

$$\frac{d}{dt}\begin{pmatrix}x_1(t)\\x_2(t)\end{pmatrix} = \frac{iky(t)}{2}\begin{pmatrix}1 & e^{-2ikt}\\-e^{2ikt} & -1\end{pmatrix}\begin{pmatrix}x_1(t)\\x_2(t)\end{pmatrix}, \qquad 0 < t < L, \tag{15.4}$$
$$x_1(0) = 1, \qquad x_2(L) = 0,$$

and

$$x_2(0) = R, \qquad x_1(L) = T. \tag{15.5}$$

Note that in (15.4), $(x_1(t), x_2(t))$ is a complex 2-dimensional vector function and that the boundary conditions have a particularly simple form (cf. (14.2)).

Let $Y(t, s)$ be the complex 2×2 fundamental matrix of (15.4). It is easily seen that it has the form

$$Y = \begin{pmatrix}a & b\\\bar{b} & \bar{a}\end{pmatrix}, \qquad |a|^2 - |b|^2 = 1, \tag{15.6}$$

where bar is complex conjugation. Thus, from

$$\begin{pmatrix}a(L,0) & b(L,0)\\\bar{b}(L,0) & \bar{a}(L,0)\end{pmatrix}\begin{pmatrix}1\\R\end{pmatrix} = \begin{pmatrix}T\\0\end{pmatrix}, \tag{15.7}$$

we deduce that

$$R = -\frac{\bar{b}}{\bar{a}}, \qquad T = \frac{1}{\bar{a}}, \qquad |T|^2 + |R|^2 = 1. \tag{15.8}$$

To emphasize dependence on parameters we may write

$$R = R(L, k; y(\cdot)), \qquad T = T(L, k; y(\cdot)).$$

We come now to the stochastic problem. From the way the index of refraction $(1 + y(t))^{1/2}$ is written, it is clear that $y(t)$ is intended to play the role of fluctuations. So we assume it is stationary and that

$$E\{y(t)\} = 0, \qquad t \geqslant 0. \tag{15.9}$$

Let $\rho(t)$ be the covariance function

$$\rho(t) = E\{y(t + s)y(s)\}/\alpha^2, \qquad \alpha^2 = E\{y^2(t)\}, \tag{15.10}$$

and let

$$l = \int_0^\infty \rho(t)\, dt, \tag{15.11}$$

assumed finite. Note that l has the dimensions of a length (t has dimension of length) and that α, the standard deviation of the fluctuations, is dimensionless. Thus the transmission coefficient, the object of principal interest, is a complex valued random variable with $|T| \leqslant 1$ and

$$T = T(L, k, l, \alpha), \tag{15.12}$$

where L, k, l and α are parameters, with α, kL and kl dimensionless.

We wish to find the probability distribution of T as a function of its parameters. This turns out to be an impossible job even when one allows $y(t)$ to be a most convenient, for calculations, process like white noise. There are at least 4 interesting asymptotic limits that one can study. We shall introduce them next and comment on their physical significance.

The first, and perhaps the simplest, is to let l be proportional to a small parameter ϵ, say $l \to l\epsilon^2$ and let $\epsilon \to 0$. Clearly

$$T(L, k, \epsilon^2 l, \alpha) \to 1 \qquad \text{as } \epsilon \downarrow 0 \tag{15.13}$$

and this is not hard to show. Next one considers the fluctuation

$$\frac{1}{\epsilon}(T(L, k, \epsilon^2 l, \alpha) - 1), \tag{15.14}$$

and one can show [60] that it tends asymptotically as $\epsilon \downarrow 0$ to a Gaussian random variable. This is the Gauss-Markov limit of Section 6. It is particularly useful when the index of refraction is not simply $(1 + \epsilon y(t))^{1/2}$, but, say, $(\zeta(t) + \epsilon y(t))^{1/2}$ where $\zeta(t)$ is deterministic (the mean profile). In this case the right-hand side of (15.13) is not 1 but the deterministic transmission coefficient.

The second case is the white noise limit (cf. (7.1)) where $\alpha \to \alpha/\epsilon$ and $l \to \epsilon^2 l$. Then $T(L, k, \epsilon^2 l, \alpha/\epsilon)$ tends (weakly) as $\epsilon \to 0$ to a random variable that can be determined in principle by solving a complicated diffusion equation on a 3-dimensional manifold. Physically, this case corresponds to large fluctuations and small correlation length relative to the wavelength of the incident wave $(2\pi/k)$ and the width L. There are no explicit results for this case as far as we know.

The third case is to take $\alpha \to \alpha\epsilon$, $L \to L/\epsilon^2$. It is easy to see that

$$T\left(\frac{L}{\epsilon^2}, k, l, \epsilon\alpha\right) = T\left(L, \frac{k}{\epsilon^2}, \epsilon^2 l, \epsilon\alpha\right) = T\left(\frac{L}{\epsilon}, \frac{k}{\epsilon}, \epsilon l, \epsilon\alpha\right). \tag{15.15}$$

Here the limit can be computed explicitly [23] as we shall describe below. It corresponds to (7.20) and (7.21). Physically we have (i) small fluctuations and (ii) either large width with k and l fixed, or large k and small l with L fixed, or large L, large k and small l as shown in (15.15). In particular, the wavelength and the correlation length are comparable while the width is large, which is a physically interesting problem.

The fourth case is $L \to L/\epsilon^2$, $k \to \epsilon k$, which leads to

$$T\left(\frac{L}{\epsilon^2}, \epsilon k, l, \alpha\right) = T\left(L, \frac{k}{\epsilon}, \epsilon^2 l, \alpha\right) = T\left(\frac{L}{\epsilon}, k, \epsilon l, \alpha\right). \tag{15.16}$$

Physically, we treat the case of fluctuations that are of order one, correlations that are of order one, low frequency and large width (first scaling in (15.16)). This limit corresponds to (7.24), (7.25) and is important in the heat conduction problem that we discuss in Section 16.

Now we return to the third scaling (15.15). According to the above discussion it is all a question of finding the asymptotic behavior of the fundamental solution process $Y^\epsilon(t, s)$ which satisfies the stochastic equation

$$\frac{dY^\epsilon(t,s)}{dt} = \frac{iky(t/\epsilon^2)}{2}\begin{pmatrix} 1 & e^{-2ikt/\epsilon^2} \\ -e^{2ikt/\epsilon^2} & -1 \end{pmatrix} Y^\epsilon(t,s), \qquad (15.17)$$

$$Y^\epsilon(s,s) = I.$$

This is a problem of the form (7.20) with the substitution (7.21), but matrix-valued. The formula (7.23) works just as well here and the problem is to solve the diffusion equation with $\bar{\mathscr{L}}$ as the diffusion operator. This is done in [23] and references therein, and the ideas are simple but there are a lot of details. We shall only give the results for the quantity T here:

$$\lim_{\epsilon\downarrow 0} E\left\{\left|T\left(\frac{L}{\epsilon^2}, k, l, \epsilon\alpha\right)\right|^2\right\} = e^{-L\beta/4}\int_{-\infty}^{\infty} e^{-t^2\beta L}\,\frac{\pi t \sinh \pi t}{\cosh^2 \pi t}\,dt,$$

$$\lim_{\epsilon\downarrow 0} E\left\{\left|T\left(\frac{L}{\epsilon^2}, k, l, \epsilon\alpha\right)\right|^4\right\} = e^{-L\beta/4}\int_{-\infty}^{\infty} e^{-t^2\beta L}\left(t^2+\frac{1}{4}\right)\frac{\pi t \sinh \pi t}{\cosh^2 \pi t}\,dt, \qquad (15.18)$$

where

$$\beta = \frac{(k\alpha)^2}{2}\int_0^{\infty} \rho(s)\cos 2ks\, ds. \qquad (15.19)$$

Note that the l of (15.11) does not enter directly but rather through the expression (15.19). Of course, the discussion concerning the scaling did not take into account finer details which necessarily enter (and affect the result).

16. HEAT CONDUCTION IN A ONE-DIMENSIONAL RANDOM MEDIUM

The formulation of this problem is as follows. We begin with (15.1) but now assume that the incident wave is not a time harmonic wave of unit amplitude and with wave number k. We assume that the incident wave at location $t = 0$ is time dependent*; specifically a stationary random function of time with decomposition

$$v_I(\tau) = \int_{-\infty}^{\infty} e^{-i\omega\tau} A(\omega)\, d\omega$$

$$\bar{A}(-\omega) = A(\omega) \qquad \text{(for } v_I \text{ to be the real-valued).} \tag{16.1}$$

Moreover

$$\langle v_I(\tau + \sigma) v_I(\tau) \rangle = \int_{-\infty}^{\infty} e^{-i\omega\sigma} \theta(\omega)\, d\omega$$

where $\theta(\omega) \geqslant 0$ is the spectral density and $\langle \cdot \rangle$ denotes average and is distinct from $E\{\cdot\}$ which involves the coefficient $y(t)$ (we assume $y(\cdot)$ and $v_I(\cdot)$ are independent).

The transmitted pulse at location $t = L$ is given by

$$v_T^L(\tau) = \int_{-\infty}^{\infty} e^{-i\omega\tau} e^{i\omega L/c} T\left(L, \frac{\omega}{c}, l, \alpha\right) d\omega$$

where c is the wave speed in vacuum (say $c = 1$). Thus

$$E\left\{\langle v_T^L(\tau + \sigma) v_T^L(\sigma) \rangle\right\} = \int_{-\infty}^{\infty} e^{-i\omega\tau} E\left\{ \left| T\left(L, \frac{\omega}{c}, l, \alpha\right) \right|^2 \right\} \theta(\omega)\, d\omega.$$

To model a heat bath on the left side we may take $\theta(\omega) \equiv 1$, in which case the quantity or primary interest is the total amount of energy transmitted (on the average) by the medium over all frequencies

$$J(L, l, \alpha) = E\{v_T^L(\tau) v_T^L(\tau)\} = 2c \int_0^{\infty} E\{|T(L, k, l, \alpha)|^2\}\, dk. \tag{16.2}$$

The problem is to find [62, 63] lim $J(L, l, \alpha)$ as $L \to \infty$. Let $L = 1/\epsilon^2$ with $\epsilon \to 0$. Then

$$J(L, l, \alpha) = 2c\epsilon \int_0^{\infty} E\left\{ \left| T\left(\frac{1}{\epsilon^2}, \epsilon \cdot \frac{k}{\epsilon}, l, \alpha\right) \right|^2 \right\} d\left(\frac{k}{\epsilon}\right)$$

$$= 2c\epsilon \int_0^{\infty} E\left\{ \left| T\left(\frac{1}{\epsilon^2}, \epsilon k, l, \alpha\right) \right|^2 \right\} dk.$$

---

* Here τ denotes time and t the space variable.

Hence

$$\lim_{L\uparrow\infty} \sqrt{L}\, J(L, l, \alpha) = \lim_{\epsilon\downarrow 0} 2c \int_0^\infty E\left\{ \left| T\left(\frac{1}{\epsilon^2}, \epsilon k, l, \alpha\right) \right|^2 \right\} dk, \tag{16.3}$$

provided the limits exist. We note that the integrand on the right is scaled in exactly the form (15.16). The problem is now the passage of the limit $\epsilon \downarrow 0$ under the integral sign. For this one needs an estimate so as to use the dominated convergence theorem. This estimate follows, essentially, from the work of Pastur and Fel'dman [61] but the details require further attention and will be given elsewhere (since they are also of independent interest). Assuming this interchange we have now

$$\lim_{L\uparrow\infty} \sqrt{L} J(L, l, \alpha) = 2c \int_0^\infty \lim_{\epsilon\downarrow 0} E\left\{ \left| T\left(\frac{1}{\epsilon^2}, \epsilon k, l, \alpha\right) \right|^2 \right\} dk. \tag{16.4}$$

We compute the ϵ limit in (16.4) exactly as in [23] but with the formula (7.26) instead of (7.23) which leads to essentially the same result up to a redefinition of β in (15.19). We find, in fact, that

$$\lim_{L\uparrow\infty} \sqrt{L} J(L, l, \alpha) = \frac{2c\pi^{3/2}}{(\alpha^2 l/2)^{1/2}} \int_0^\infty \left(t^2 + \frac{1}{4}\right)^{-1/2} \frac{t \sinh \pi t}{\cosh^2 \pi t}\, dt. \tag{16.5}$$

The physical significance of this result is the following. The integrated (over frequencies) or total transmittance of the medium decreases as $L \to \infty$ (all other parameters being fixed) like $1/\sqrt{L}$. Moreover, most of the transmission occurs in the neighborhood of frequencies proportional to $1/\sqrt{L}$, i.e., low frequencies. The medium acts asymptotically like a low pass filter. For further information on the physics of the problem we refer to [62, 63].

REFERENCES

1. W. Feller, *An Introduction to Probability and its Applications*, Wiley, New York, 1966.
2. J. L. Doob, *Stochastic Processes*, Wiley, New York, 1953.
3. I. I. Gihman and A. V. Skorohod, *Stochastic Differential Equations*, Springer, Berlin–Heidelberg–New York, 1972.

4. H. P. McKean, Jr., *Stochastic Integrals*, Academic Press, New York, 1968.
5. E. B. Dynkin, *Markov Process I, II*, Springer, Berlin–Heidelberg–New York, 1965.
6. D. Stroock and S. R. S. Varadhan, "Diffusion processes with continuous coefficients I, II," *Comm. Pure Appl. Math.*, **22** (1969), 345–400, 479–530.
7. D. Stroock, "Diffusion processes associated with Lévy generators," *Z. Wahrscheinlichkeitstheorie und Verw. Gebiete*, **32** (1975), 209–244.
8. D. Stroock and S. R. S. Varadhan, "Diffusion processes with boundary conditions," *Comm. Pure Appl. Math.*, **24** (1971), 147–225.
9. I. I. Gihman and A. V. Skorohod, *Stochastic Processes I*, Springer, Berlin–Heidelberg–New York, 1974.
10. T. Kato, *Perturbation of Linear Operators*, Springer, Berlin–Heidelberg–New York, 1966.
11. T. G. Kurtz, "Extensions of Trotter's operator semigroup approximation theorems," *J. Funct. Anal.*, **3** (1969), 354–375.
12. J. Neveu, "Potentiel Markovien récurrent des chaines de Harris," *Ann. Inst. Fourier, Grenoble*, **22** (1972), 85–130.
13. S. Orey, *Limit Theorems for Markov Chain Transition Probabilities*, Van Nostrand, Princeton, N.J., 1969.
14. V. Beneš, "Finite regular invariant measures for Feller processes," *J. Appl. Probability*, **5** (1968), 203–209.
15. M. Rosenblatt, *Markov Processes, Structure and Asymptotic Behavior*, Springer-Verlag, Berlin, 1971.
16. ———, Dependence and asymptotic independence for random processes, this study.
17. N. N. Bogoliubov and Yu. Mitropolsky, *Asymptotic Methods in Nonlinear Mechanics*, Gordon and Breach, New York, 1961.
18. Yu. Mitropolski, "Averaging methods in nonlinear mechanics," *Int. J. Non-Linear Mech.*, **2**, 69–96.
19. J. Hale, *Ordinary Differential Equations*, Wiley, New York, 1969.
20. T. G. Kurtz, Applications of an abstract perturbation theorem to ordinary differential equations (to appear).
21. R. Z. Khasminskii, The averaging principle for parabolic and elliptic differential equations and Markov processes with small diffusion," *Theor. Prob. Appl.*, **8** (1963), 1–21.
22. E. B. Davies, "Asymptotic analysis of some abstract evolution equations," *J. Funct. Anal.* (to appear).
23. W. Kohler and G. Papanicolaou, "Power statistics for waves in one dimension and comparison with radiative transport theory I, II," *J. Math. Phys.*, **14** (1973), 1733–1745, **15** (1974), 2186–2197.

24. K. Case, "A general perturbation method for quantum mechanical problems," *Suppl. Progress in Th. Phys.*, **37–38** (1966), 1–20.
25. M. I. Freidlin, "Dirichlet's problem for an equation with periodic coefficients depending on a small parameter," *Theory Prob. Appl.*, **9** (1964), 121–125.
26. R. Z. Khasminskii, "On the principle of averaging for Itô's stochastic differential equations," *Kybernetica (Prague)*, **4** (1968), 260.
27. A. Bensoussan, J. L. Lions, and G. C. Papanicolaou, "Homogenization and ergodic theory," *Proc. of Conf. in Warsaw, Poland*, April 1976 (to appear).
28. ———, *Asymptotic Problems in Periodic Structures*, North-Holland, Amsterdam, 1978.
29. I. Babuška, Several reports from the Department of Mathematics, University of Maryland.
30. R. Z. Khasminskii, "On the stochastic processes defined by differential equations with a small parameter," *Theor. Prob. Appl.*, **11** (1966), 211–222.
31. B. White, "The effects of a rapidly fluctuating environment on systems of interacting species," *SIAM J. Appl. Math.*, **32** (1977), 666–693.
32. G. C. Papanicolaou, "Asymptotic analysis of transport processes," *Bull. Amer. Math. Soc.*, **81** (1975), 330–392.
33. M. F. Norman, *Markov Processes and Learning Theory*, Academic Press, New York, 1972.
34. R. Z. Khasminskii, "A limit theorem for solutions of differential equations with a random right hand side," *Theor. Prob. Appl.*, **11** (1966), 390–406.
35. G. C. Papanicolaou and W. Kohler, "Asymptotic theory of mixing stochastic ordinary differential equations," *Comm. Pure Appl. Math.*, **27** (1974), 641–668.
36. G. C. Papanicolaou, "Some probabilistic problems and methods in singular perturbations," *Rocky Mountain J. of Math.*, **6** (1976), 653–674.
37. T. G. Kurtz, "A limit theorem for perturbed operator semigroups with applications to random evolution," *J. Funct. Anal.*, **12** (1973), 55–67.
38. R. L. Stratonovich, *Topics in the Theory of Random Noise*, vols. I, II, Gordon and Breach, New York, 1963.
39. E. B. Davies, "Markovian master equations I," *Comm. Math. Phys.*, **39** (1974), 91–110.
40. G. C. Papanicolaou and S. R. S. Varadhan, "A limit theorem with strong mixing in Banach space," *Comm. Pure Appl. Math.*, **26** (1973), 497–524.
41. G. C. Papanicolaou and W. Kohler, "Asymptotic analysis of deterministic and stochastic equations with rapidly varying components," *Comm. Math. Phys.*, **45** (1975), 217–232.

42. A. Khinchine, *Asymptotische Gesetze der Wahrscheinlichkeitsrechnung*, Springer, Berlin, 1933.

43. A. V. Skorohod, *Studies in the Theory of Random Processes*, Addison-Wesley, Reading, Mass., 1964.

44. F. John, "On integration of parabolic equations by difference methods I," *Comm. Pure Appl. Math.*, **5** (1952), 155–211.

45. H. Kushner, "Probabilistic methods for finite difference approximations to degenerate elliptic and parabolic equations with Neumann and Dirichlet boundary conditions," *J. Math. Anal. Appl.*, **53** (1976), 644–668.

46. R. Z. Khasminskii, *Stability of systems of differential equations under random perturbations of their parameters*, Nauka, Moscow, 1969.

47. M. Pinsky, "Stochastic stability and the Dirichlet problem," *Comm. Pure Appl. Math.*, **27** (1974), 311–350.

48. G. Blankenship and G. C. Papanicolaou, "Stability and control of systems with wide-band noise disturbances I, II," *SIAM J. Appl. Math.*, **34** (1978), 423–476.

49. A. Friedman, *Stochastic Differential Equations and Applications I, II*, Academic Press, New York, 1975.

50. D. Ludwig, "On persistence in dynamical systems," *SIAM Rev.*, **17** (1975), 605–640.

51. M. G. Krein and M. A. Rutman, "Linear operators leaving invariant a cone in a Banach space," *Uspehi Mat. Nauk*, **3**, No. 1 (23) (1948), 3–95, and *Amer. Math. Soc. Transl.*, **26** (1950).

52. A. Bensoussan, J. L. Lions, and G. C. Papanicolaou, "Boundary layers and homogenization of transport processes," *Journal of RIMS*, Kyoto, Japan (to appear).

53. N. Levinson, "The first boundary value problem for $\epsilon\Delta u + Au_x + Bu_y + C = D$ for small ϵ," *Ann. of Math.*, **5** (1950), 428–445.

54. C. Holland, "Singular perturbation problems using probabilistic methods," *Rocky Mountain J. of Math.*, **6** (1976), 585–590.

55. A. D. Ventsel and M. I. Freidlin, "On small random perturbations of dynamical systems," *Russian Math. Surveys*, **25** (1970), 1–55.

56. A. Bensoussan and J. L. Lions, "Diffusion processes in bounded domains and singular perturbation problems for variational inequalities with Neumann boundary conditions," in *Probabilistic Methods in Differential Equations*, Lecture Notes in Math., #451, Springer, New York, 1975.

57. C. Bardos, "Problèmes aux limites pour les équations aux dérivées partielles du premier ordre à coefficients réels; théorèmes d'approximation; application à l'équation de transport," *Ann. Sci. École Norm. Sup.*, **3** (1970), 185–233.

58. B. Matowsky and Z. Schuss, "The exit problem for randomly perturbed dynamical systems," *SIAM J. Appl. Math.*, **33** (1977), 365–382.

59. G. C. Papanicolaou, D. W. Stroock, and S. R. S. Varadhan, "Martingale approach to some limit theorems," in *Statistical Mechanics Dynamical Systems and the Duke Turbulence Conference*, by D. Ruelle, Duke Univ. Math. Series, Vol. 3, Durham, N.C., 1977.

60. J. Bazer (to appear).

61. L. A. Pastur and E. P. Fel'dman, "Wave transmittance for a thick layer of a randomly inhomogeneous medium," *Zh. Eksp. Teor. Fiz.*, **67** (1974), 487–489 (*Soviet Phys. JETP*, **40** (1975), 241–243).

62. A. J. O'Connor, "A central limit theorem for the disordered harmonic chain," *Comm. Math. Phys.*, **45** (1975), 63–77.

63. A. Casher and J. L. Lebowitz, "Heat flow in regular and disordered harmonic chains," *J. Math. Phys.*, **12** (1971), 1701–1711.

SOME MATHEMATICAL PROBLEMS IN STATISTICAL MECHANICS

M. Kac

1. STATISTICAL MECHANICS OF A ONE-DIMENSIONAL GAS

1. This is the English original of the text which formed the basis of a series of lectures I gave some years ago at the University of Montreal as the first holder of the Aisenstadt chair. The French version (which also included a brief chapter on the problem of the disordered chain) was published by the University of Montreal Press (Les Presses de l'Université de Montréal).

Part 1 is a somewhat enlarged version of the major part of my Gibbs lecture delivered in January 1967 in Houston, Texas, and the second (quite independent of the first) an introduction to the problematics of the theory of phase transitions using the spherical model as a convenient (though unrealistic) vehicle.

In spite of the adjective "statistical," Statistical Mechanics as it pertains to systems in *thermal equilibrium* has relatively little to do with Probability Theory. Only when it comes to the problem of *approach* to equilibrium does the use of probabilistic concepts (notably those related to ergodic theory) become essential.

The inclusion of this paper in this study may nevertheless be justified by intrinsic interest of the problems related to phase transitions and by the hope that these problems will ultimately influence the development of Probability Theory.

2. Assuming that one has at least a vague familiarity with concepts and terminology of classical thermodynamics let me review briefly some of its most important statements as they pertain to the simplest systems, i.e., gases.

(a) A state of a gas in equilibrium is completely specified by its volume V and its pressure p.

(b) For each gas G in equilibrium there is a function $f_G(p, V)$ called its *absolute temperature* such that if two gases G_1 and G_2 in respective states (p_i, V_i) are brought into thermal contact with each other they remain in equilibrium if and only if

$$f_{G_1}(p_1, V_1) = f_{G_2}(p_2, V_2). \tag{2.1}$$

The equation

$$T = f_G(p, V) \tag{2.2}$$

which determines the absolute temperature of a gas in equilibrium once its pressure p and volume V are known is called the *equation of state.*

It is customary to write the equation of state in the form

$$p = F(V, T), \tag{2.3}$$

and it follows from the Second Law* that

$$\frac{\partial p}{\partial V} \leqslant 0, \tag{2.4}$$

i.e., the isotherm (2.3) is monotonic.

---

* More precisely from the Second Law as it pertains to *irreversible* changes. See Uhlenbeck and Ford [1], Chapter I.

(c) There is a function $U(V, T)$ called *internal energy* of the gas such that, in thermal isolation, the change in U is equal to the work done by, or on, the gas.

In particular, if the gas, in thermal isolation, undergoes a *reversible** volume change dV then

$$dU + p\,dV = 0. \tag{2.5}$$

If a similar reversible change in volume takes place in thermal contact with other bodies then $dU + p\,dV$ is not zero. In fact, $dU + p\,dV$ is the energy given to or taken away from the system by the other bodies; and since the only form of energy (apart from mechanical) which one considers is thermal, it is in keeping with the law of conservation of energy to *define* $dU + p\,dV$ as the *quantity of heat* δQ added to (or subtracted from) the system:

$$\delta Q = dU + p\,dV. \tag{2.6}$$

(d) It follows from that part of the Second Law which pertains to reversible changes that

$$\frac{\delta Q}{T} = \frac{dU}{T} + \frac{p}{T}\,dV \tag{2.7}$$

is an *exact differential.*

Thus there is a function $S(V, T)$ (defined up to an additive constant) such that

$$dS = \frac{\delta Q}{T}. \tag{2.8}$$

This function is the famed *entropy*.

(e) It is convenient to introduce another function $\Psi(V, T)$, defined by the formula

$$\Psi = U - ST \tag{2.9}$$

and called the *free energy* (actually the Helmholtz free energy).

* A reversible process is one in which the gas while undergoing a change is at each instant of time in thermal equilibrium. This represents, of course, a far-reaching idealization and requires, e.g., that all such changes take place infinitely slowly.

One then verifies immediately that

$$p = -\frac{\partial \Psi}{\partial V}. \tag{2.10}$$

I apologize for such a sketchy and inadequate summary of the elements of thermodynamics.

I certainly do not want to leave the impression that the subject is trivial or that I have said all there is to be said about it with reference to gases. Thermodynamics is actually quite subtle, and its basic concepts of equilibrium and of temperature far from obvious or superficial.

I am simply using my prerogative as a mathematician to *postulate* what I need (and what I *know* to be correct!) and to take as primitive notions those which I find difficult or awkward to define.

3. Thermodynamics provides us only with *relations* between *elastic* properties of a gas as described by its equation of state and its *caloric* properties as described by the internal energy function $U(V, T)$.

Thus if one knows, e.g., that the equation of state is the Boyle-Charles law

$$p = \frac{RT}{V}, \tag{3.1}$$

it *follows* (from the fact that $\delta Q/T$ is an exact differential) that

$$\frac{\partial U}{\partial V} = 0, \tag{3.2}$$

i.e., that U is a function of temperature alone.

But this is as far as Thermodynamics can go. To actually *derive* the equation of state and the functional form of U one must bring into the theory considerations concerning structure of matter, and it is at this point that Statistical Mechanics comes in.

4. A monoatomic gas confined to a container $\mathscr{V}$ of volume V is viewed as a mechanical system of N particles of mass m with total energy

$$E = \frac{1}{2m}\sum_{1}^{N} \vec{p}_k^{\,2} + \phi(\vec{r}_1, \ldots, \vec{r}_N) + \sum_{k=1}^{N} u(\vec{r}_k). \tag{4.1}$$

Here $\vec{r}_k$ and $\vec{p}_k$ are respectively the position vector and momentum of the kth particle, ϕ the interaction potential, and $u(\vec{r})$ the "wall potential" whose purpose is to keep the particles in $\mathscr{V}$.

One assumes that

$$u(\vec{r}) = \begin{cases} 0, & \vec{r} \in \mathscr{V} \\ \infty, & \vec{r} \notin \mathscr{V} \end{cases} \tag{4.2}$$

and that

$$\phi(\vec{r}_1, \ldots, \vec{r}_N) = \sum_{1 \le i < j \le N} \phi(|\vec{r}_i - \vec{r}_j|), \tag{4.3}$$

i.e., that particles interact through two-body central forces.

The interparticle potential $\phi(r)$ is assumed to be of the form shown on the graph in Figure 1, where in particular

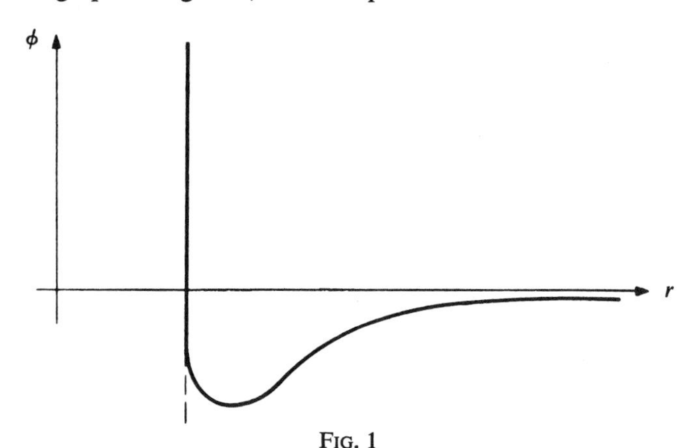

FIG. 1

$$\phi(r) = +\infty, \qquad 0 \leqslant r < \delta, \qquad \text{(hard core assumption)}, \tag{4.4}$$

i.e., no two particles can come closer than δ (they are like hard impenetrable spheres of diameter δ).

Within this framework the fundamental postulate of classical statistical mechanics can be put as follows:

The statement that a (monoatomic) gas confined to a container $\mathscr{V}$ is in equilibrium at absolute temperature T is to be interpreted to mean that the probability of finding the gas in a state within the differential volume $d\Gamma$ $(= d\vec{p}_1\, d\vec{p}_2 \cdots d\vec{p}_N\, d\vec{r}_1\, d\vec{r}_2 \cdots d\vec{r}_N)$ of the point $\vec{p}_1, \ldots, \vec{p}_n, \vec{r}_1, \ldots, \vec{r}_n$ of the $6N$-dimensional phase space Γ is

$$D(\vec{p}_1, \ldots, \vec{p}_N; \vec{r}_1, \ldots, \vec{r}_N)\, d\Gamma = \frac{e^{-(E/kT)}\, d\Gamma}{\int e^{-(E/kT)}\, d\Gamma}. \tag{4.5}$$

In this formula k is a certain constant (known as the Boltzmann constant) which fixes the temperature scale.

Internal energy U is defined as the average, with respect to D, of the energy E of the system, i.e.,

$$U = \langle E \rangle = \frac{\int E e^{-(E/kT)}\, d\Gamma}{\int e^{-(E/kT)}\, d\Gamma} \tag{4.6}$$

and entropy S is defined as the average of $-k \log D$ (again with respect to D), i.e.,

$$S = -k\langle \log D \rangle = -k \int D \log D\, d\Gamma. \tag{4.7}$$

Having defined energy and entropy we define free energy by the formula (2.9), i.e.,

$$\Psi = U - TS \tag{4.8}$$

and finally pressure p is given by formula (2.10)

$$p = -\frac{\partial \Psi}{\partial V}. \tag{4.9}$$

In this way not only is the fundamental identity of thermodynamics

$$dS = \frac{dU}{T} + \frac{p\, dV}{T} \tag{4.10}$$

preserved but a *definite recipe* is also provided which enables one, in principle, to *derive* the equation of state (4.9) and the dependence of internal energy U on volume and temperature.

The postulate embodied in (4.5) as well as the statistical interpretation of entropy contained in (4.7) are *not* proved. They can only be made highly plausible by a combination of mathematical and physical arguments which involve, among others, reference to the ergodic theorem.

A clear and detailed justification can be found in [1].

Of considerable interest is the justification given by Gibbs himself (in his famous book *Elementary Principles in Statistical Mechanics, Developed with Special Reference to the Rational Foundation of Thermodynamics*).

Gibbs takes a highly formal point of view, and he refers to U and S as defined by formulas (4.6) and (4.7) only as "thermodynamic analogies." This extreme caution was not merely a manifestation of Gibbs' strict logical attitude. At the time he wrote his book statistico-mechanical calculations applied to specific heats of diatomic gases led to results which were at variance with experiment. Worse yet, in the theory of black body radiation statistical mechanics led to a divergent result for the total energy (so-called ultraviolet catastrophe, since divergence came from the short wavelength part of the spectrum).

"Difficulties of this kind," wrote Gibbs in the introduction of his book, "have deterred the author from attempting to explain the mysteries of nature, and have forced him to be contented with the more obvious propositions relating to the statistical branch of mechanics. Here, there can be no mistake in regard to the agreement of the hypotheses with the facts of nature, for nothing is assumed in that respect. The only error into which one can fall is the want of agreement between the premises and the conclusions, and this, with care, one may hope, in the main, to avoid."

These difficulties were soon resolved by quantum theory, though Gibbs unfortunately did not live to witness this great new development.

5. Long before one becomes involved with quantum difficulties one must face the problem of whether the Gibbs formalism as described in the preceding section is capable of explaining so simple and so widespread a phenomenon as condensation of gases.

It is well known that every gas, if cooled to a sufficiently low temperature, will liquify when subjected to a sufficiently high pressure. On the other hand, if the temperature is not sufficiently low, no pressure will produce liquefication though ultimately solidification will be achieved.

Disregarding the problem of solidification the experimental facts concerning condensation (liquefication) are as follows:

There is a critical temperature T_c such that for $T > T_c$ the isotherm is smooth and looks as shown in Figure 2.

For $T < T_c$ the isotherm is composed of three smooth pieces, the middle one being a straight line segment parallel to the V-axis.* (See Figure 3.)

We are thus faced with a dramatic change in analytic behavior of isotherms which must be understood on the basis of the dynamical and statistical assumptions of the preceding sections.

Let us first note that (with $\beta = 1/kT$)

$$\int e^{-(E/kT)}\,d\Gamma = \int e^{-\beta E}\,d\Gamma$$

$$= \left(\frac{2\pi m}{\beta}\right)^{3N/2} \int_{\mathscr{V}} \cdots \int_{\mathscr{V}}$$

$$\times \exp\left\{-\beta \sum_{1\le i<j\le N} \phi(|\vec{r}_i - \vec{r}_j|)\right\} d\vec{r}_1 \cdots d\vec{r}_N \tag{5.1}$$

and let us set

$$Q_N(\mathscr{V}, \beta) = \int_{\mathscr{V}} \cdots \int_{\mathscr{V}} \exp\left\{-\beta \sum_{1<i\le j\le N} \phi(|\vec{r}_i - \vec{r}_j|)\right\} d\vec{r}_1 \cdots d\vec{r}_N. \tag{5.2}$$

It follows easily that

$$U = \frac{3N}{2}\frac{1}{\beta} - \frac{\partial}{\partial\beta} \log Q_N(\mathscr{V}; \beta) \tag{5.3}$$

---

* As the volume is decreased from V_2 to V_1 the pressure remains constant while all of the gas becomes liquefied.

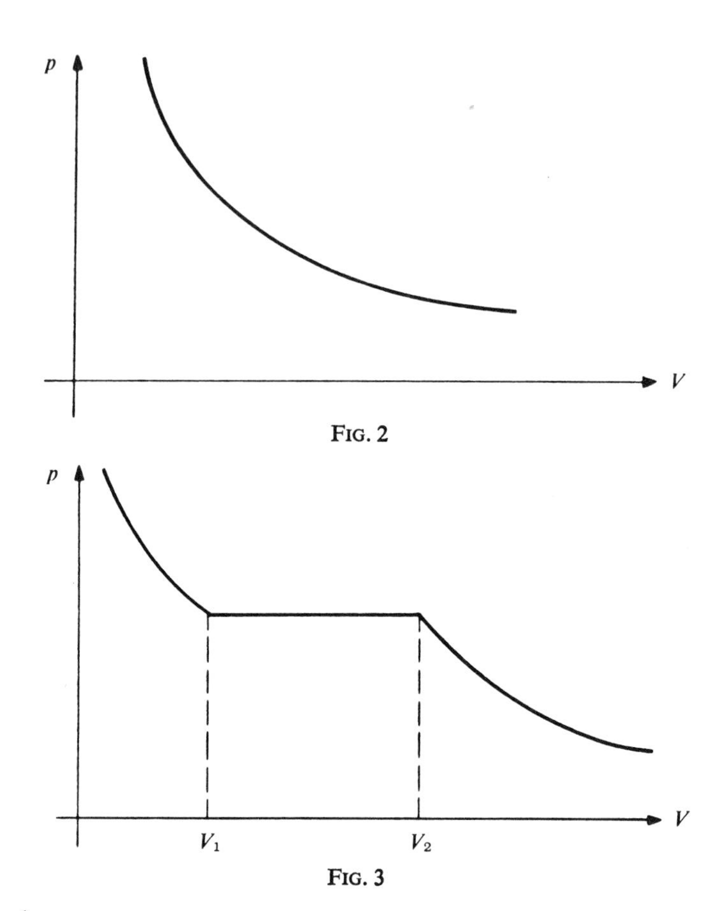

FIG. 3

and

$$\Psi = -\frac{3Nk}{2}\log\frac{2\pi m}{\beta} - \frac{1}{\beta}\log Q_N(\mathscr{V};\beta). \tag{5.4}$$

Thus

$$p = -\frac{\partial\Psi}{\partial V} = \frac{1}{\beta}\frac{\partial}{\partial V}\log Q_N(\mathscr{V};\beta). \tag{5.5}$$

A glance at (5.2) will convince one that as long as N and V are finite p will be an analytic function of V and β.

We must therefore look to a *limiting process* for an explanation of the change in the analytic behavior of the isotherms.

The limiting process consists in letting

$$N \to \infty, \qquad V \to \infty \tag{5.6}$$

but in keeping the specific volume

$$v = \frac{V}{N} \tag{5.7}$$

fixed. (It should also be understood that $V \to \infty$ means that "all dimensions" of $\mathscr{V}$ approach infinity.)

This limit will be referred to as the *thermodynamic limit.*

We rewrite (5.5) in the form

$$p = \frac{1}{\beta}\frac{\partial}{\partial v}\left\{\frac{1}{N}\log Q_N(\mathscr{V};\beta)\right\} \tag{5.8}$$

and all would be well if the limit of

$$\frac{1}{N}\log Q_N(\mathscr{V};\beta)$$

existed as $N \to \infty$, $V \to \infty$, $V/N = v$. Unfortunately, as the example of the ideal gas

$$Q_N(\mathscr{V};\beta) = V^N \tag{5.9}$$

already shows this is not the case. However, this failure for the ideal gas is of a trivial nature since

$$\frac{1}{N}\log Q_N(\mathscr{V};\beta) = \log N + \log v \tag{5.10}$$

and the trouble is caused by the term $\log N$ which being independent of v disappears when the derivative with respect to v is taken. This suggests replacing, in general, $Q_N(\mathscr{V};\beta)$ by

$$\frac{Q_N(\mathscr{V};\beta)^*}{N!}$$

---

* The choice of $N!$ is to a large extent arbitrary. Anything which is asymptotically $(cN)^N$, with c a constant independent of v will do.

so that formula (5.8) becomes

$$p = \frac{1}{\beta} \frac{\partial}{\partial v} \left\{ \frac{1}{N} \log \frac{Q_N(\mathscr{V}; \beta)}{N!} \right\}. \tag{5.11}$$

It has been shown by Ruelle (see Fisher [2] where a review, an extension, and a list of references is given) that in the thermodynamic limit

$$\frac{1}{N} \log \frac{Q_n(\mathscr{V}; \beta)}{N!}$$

approaches a limit which is denoted by $-\beta f(v; \beta)$ and that, in the same limit, p approaches $-\partial f/\partial v$.

Ruelle also proved that p as a function of v is nonincreasing. We may therefore take

$$p = -\frac{\partial f}{\partial v} \tag{5.12}$$

as the limiting equation of state and the problem is whether, below a certain critical temperature, the graph of (5.12) looks as shown in Figure 4.

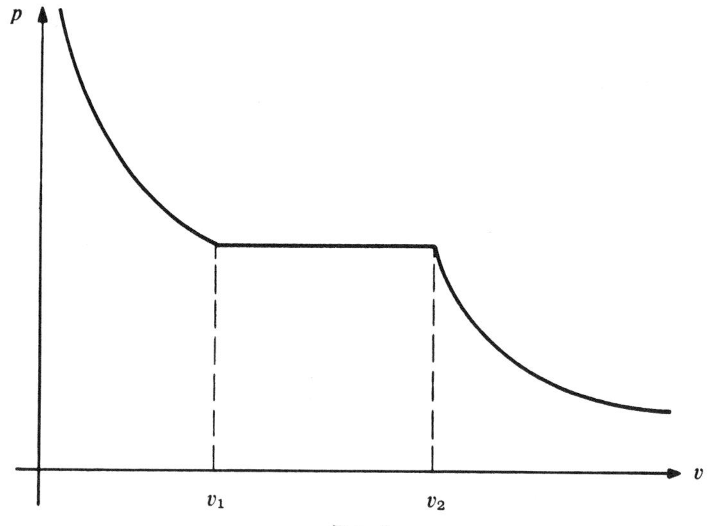

FIG. 4

6. Before we go on let me present the problem from a different point of view.

For given N and V (and as usual, since we are dealing with systems in equilibrium for a given T) we can ask what is the probability

$$q_k(\vec{r}_1, \vec{r}_2, \ldots, \vec{r}_k; \mathscr{V}, N)\, d\vec{r}_1, \ldots, d\vec{r}_k$$

that specified particles will be found at $\vec{r}_1, \ldots, \vec{r}_k$ within volume elements $d\vec{r}_1, \ldots, d\vec{r}_k$.

It follows at once from the basic postulate (4.5) that

$$q_k(\vec{r}_1, \vec{r}_2, \ldots, \vec{r}; \mathscr{V}, N)$$
$$= \frac{\int_{\mathscr{V}} \cdots \int_{\mathscr{V}} \exp\left\{-\beta \sum_{1 \le i < j \le N} \phi(|\vec{r}_i - \vec{r}_j|)\right\} d\vec{r}_{k+1}, \ldots, d\vec{r}_N}{Q_n(\mathscr{V}; \beta)} \tag{6.1}$$

but it is customary to introduce closely related functions $n_k(\vec{r}_1, \vec{r}_2, \ldots, \vec{r}_k; \mathscr{V}, N)$ defined by the formula

$$n_k(\vec{r}_1, \vec{r}_2, \ldots, \vec{r}_k; \mathscr{V}, N) = \frac{N!}{(N-k)!} q_k(\vec{r}_1, \ldots, \vec{r}_k; \mathscr{V}, N). \tag{6.2}$$

The advantage of the n_k's over the q_k's is due to the fact that in the thermodynamic limit the q_k's approach 0 while the n_k's approach nontrivial limits

$$\lim_{\substack{N \to \infty \\ V \to \infty \\ V/N = v}} n_k(\vec{r}_1, \vec{r}_2, \ldots, \vec{r}_k; \mathscr{V}, N) = \bar{n}_k(\vec{r}_1, \ldots, \vec{r}_k; v). \tag{6.3}$$

It should be understood that in taking the limit one is careful to keep all the k-particles "well within $\mathscr{V}$", i.e., let the distances between the particles and the walls of $\mathscr{V}$ approach ∞.

With this precaution one has

$$\bar{n}_1(\vec{r}; v) = \frac{1}{v} \tag{6.4}$$

$$\bar{n}_2(\vec{r}_1, \vec{r}_2, v) = \bar{n}_2(|\vec{r}_1 - \vec{r}_2|; v) \tag{6.5}$$

and in general $\bar{n}_k(\vec{r}_1, \vec{r}_2, \ldots, \vec{r}_k; v)$ depends only on the mutual distances $|\vec{r}_i - \vec{r}_j|$ between the particles.

Suppose now that the temperature is below the critical and $v_1 < v < v_2$. Under these circumstances physics tells us that two distinct phases coexist, i.e., we have some sort of a mixture of or species of specific volume v_2 (vapor) and of a species of specific volume v_1 (liquid).

In the absence of an external force field (except, of course, wall forces) the two phases are not separated in a clear-cut way by an interface but the picture is rather that of a "blob" of liquid—infinite in extent in the thermodynamic limit—placed at random in the surrounding gas.

If this picture is correct we can say that

$$\bar{n}_k(\vec{r}_1, \ldots, \vec{r}_k; v) = \begin{cases} \bar{n}_k(\vec{r}_1, \ldots, \vec{r}_k; v_1) & \text{if } \vec{r}_1 \text{ is in the liquid} \\ \bar{n}_k(\vec{r}_1, \ldots, \vec{r}_k; v_2) & \text{if } \vec{r}_1 \text{ is the gas} \end{cases} \tag{6.6}$$

If we wrote

$$v = \xi_1 v_1 + \xi_2 v_2 \tag{6.7}$$

with

$$\xi_1 + \xi_2 = 1 \tag{6.8}$$

it is clear that the probability that $\vec{r}_1$ is in the liquid is

$$\frac{\xi_1 v_1}{v}$$

and the probability that $\vec{r}_1$ is in the gas is $(\xi_2 v_2)/v$.

Thus if $\vec{r}_1$ is *not* specified we should expect that $n_k(\vec{r}_1, \ldots, \vec{r}_k; v)$ is the average

$$\bar{n}_k(\vec{r}_1, \ldots, \vec{r}_k; v) = \frac{\xi_1 v_1}{v} \bar{n}_k(\vec{r}_1, \ldots, \vec{r}_k; v_1) + \frac{\xi_2 v_2}{v} \bar{n}_k(\vec{r}_1, \ldots, \vec{r}_k; v_2). \tag{6.9}$$

To see how these considerations are related to the discussion of the equation of state, I must refer to a famous formula known as the Virial Theorem.

The Virial Theorem connects the pressure with $\bar{n}_2(r; v)$ ($r = |\vec{r}_1 - \vec{r}_2|$) and is embodied in the equation

$$p = \frac{kT}{v} + \frac{2}{3}\pi\delta^3 \bar{n}_2(\delta^+; v) - \frac{2}{3}\pi \int_\delta^\infty dr\, r^3 \frac{\partial \varphi}{dr} \bar{n}_2(r; v). \tag{6.10}$$

This formula has first been derived (in a somewhat different form) by Clausius from kinetic theory, and he used implicitly the ergodic theorem (this was before Boltzmann, who first introduced the word and the concept though not quite in agreement with modern usage). It can be proved rigorously* (under mild assumptions on ϕ) from the definitions of pressure (see (5.12)) and n_2.

It should be clear that

$$\bar{n}_2(\delta^+; v) = \lim_{r \downarrow \delta} n_2(r; v)$$

while

$$\bar{n}_2(\delta^-; v) = \lim_{r \uparrow \delta} n_2(r; v) = 0$$

(since for $r < \delta$, $\bar{n}_2(r; v) \equiv 0$).

It follows easily from the Virial Theorem that if

$$\bar{n}_2(r; v) = \frac{\xi_1 v_1}{v} \bar{n}(r; v_1) + \frac{\xi_2 v_2}{v} \bar{n}(r; v_2) \tag{6.11}$$

then

$$p(v) = \frac{\xi_1 v_1}{v} p(v_1) + \frac{\xi_2 v_2}{v} p(v_2) \tag{6.12}$$

so that the pressure is a *linear function of volume* in the region of coexistence of two phases.

If $p(v_1) = p(v_2)$, then (6.12) implies that $p(v) = p(v_1) = p(v_2)$ for all v between v_1 and v_2 (i.e., the pressure is constant for $v_1 \leqslant v \leqslant v_2$), and although this follows already from monotonicity of pressure proved by Ruelle one cannot fail to be impressed by the intimate connection which exists between the constancy of pressure and the linear combination property of n_2.

What mathematical mechanism could account for the sudden emergence of two coexistent (but geometrically separated) phases as expressed by the property (6.9) (and especially (6.11)) of distribution functions?

---

* See, e.g., H. S. Green [3].

To this day this question remains largely unanswered. Some inkling of its depth can be gathered from the fact that one-dimensional systems with short-range attraction forces cannot exhibit the phenomenon of condensation. More precisely, it was proved by van Hove [4] that the isotherms of one-dimensional systems

$$\phi(r) = 0, \qquad r > a,$$

are analytic for all temperatures.

In the next few sections I shall however present a one-dimensional model, with an attractive potential which depends on a parameter γ, and which in the limit $\gamma \to 0$ will exhibit condensation.

7. The model consists of N points in the interval $(0, L)$ interacting through the potential ϕ given by the formula

$$\phi(r) = \begin{cases} +\infty, & 0 \leqslant r < \delta, \\ -\alpha\gamma e^{-\gamma r}, & r > \delta. \end{cases} \tag{7.1}$$

Here $\alpha > 0, \gamma$ has the dimensions of inverse length and $\alpha\gamma$ the dimensions of energy. The partition function divided by $N!$ is given by the formula

$$\frac{Q_N(L)}{N!} = \frac{1}{N!} \int_0^L \cdots \int_0^L \prod_{1 \le i < j \le N} S_\delta(|t_j - t_i|)$$

$$\times \exp\left(\nu\gamma \sum_{1 \le i < j \le N} e^{-\gamma|t_j - t_i|}\right) d\Omega \tag{7.2}$$

where

$$\nu = \frac{\alpha}{kT} \tag{7.3}$$

$$d\Omega = dt_1 \ldots dt_N \tag{7.4}$$

and

$$S_\delta(r) = \begin{cases} 0, & 0 < r < \delta, \\ 1, & r > \delta. \end{cases} \tag{7.5}$$

It is clear that the product of the $S_\delta(|t_j - t_i|)$ accounts for the hard core by "killing" the integrand whenever at least two points come closer than δ to each other.

Noting that for $0 < t_1 < t_2 < \cdots < t_N$

$$\prod_{1 \le i < j \le N} S_\delta(|t_j - t_i|) = \prod_{i=1}^{N-1} S_\delta(t_{i+1} - t_i)$$

(which is the principal simplification resulting from being in one dimension) we have

$$\frac{Q_N(L)}{N!} = \exp\left(-\frac{N\nu\gamma}{2}\right) \int \cdots \int_{0<t_1<t_2<\ldots<t_N<L} \prod_{i=1}^{N-1} S_\delta(t_{i+1} - t_i)$$

$$\times \exp\left(\frac{\nu\gamma}{2} \sum_{i,j=1}^{N} e^{-\gamma|t_j - t_i|}\right) d\Omega. \tag{7.6}$$

By the use of the weird-looking identity*

$$\exp\left(\frac{\nu\gamma}{2} \sum_{i,j=1}^{N} e^{-\gamma|t_j - t_i|}\right)$$

$$= \int_{-\infty}^{\infty} \cdots \int_{-\infty}^{\infty} \exp\left(\nu\gamma \sum_{1}^{N} x_i\right) W(x_i)$$

$$\times \prod_{i=1}^{N-1} P_\gamma(x_i \mid x_{i+1}; t_{i+1} - t_i)\, dx_1 \cdots dx_N \tag{7.7}$$

---

* The identity (7.7) is not really as strange as it looks. If one knows the rudiments of the theory of stochastic processes one should recognize that (7.7) is an almost immediate consequence of the following. Let $X(t)$ be the stationary, Gaussian, Markov process with mean 0 and covariance $\exp(-\gamma|t|)$ then

$$E\left\{\exp\left(\nu\gamma \sum_{1}^{N} X(t_i)\right)\right\} = \exp\left(\frac{\nu\gamma}{2} \sum_{i,j=1}^{N} e^{-\gamma|t_j - t_i|}\right)$$

where $E\{\cdots\}$ denotes as usual the mathematical expectation of the quantity inside the braces.

For $t_1 < t_2 < \cdots < t_N$ the product

$$W(x_1) \prod_{i=1}^{N-1} P_\gamma(x_i \mid x_{i+1}; t_{i+1} - t_i)$$

is simply the density of the joint distribution of the random variables $X(t_1), X(t_2), \ldots, X(t_N)$.

However, the reader need only believe or verify for himself directly formula (7.7). He need not be familiar with stochastic processes to understand this paper.

where

$$W(x) = \frac{1}{\sqrt{2\pi}} \exp\left(-\frac{x^2}{2}\right) \tag{7.8}$$

and

$$P_\gamma(x|y;t) = \frac{\exp\{-[(y - xe^{-\gamma t})^2/2(1 - e^{-2\gamma t})]}{\sqrt{2\pi(1 - e^{-2\gamma t})}} \tag{7.9}$$

and going over to Laplace transforms one obtains without much trouble that

$$\int_0^\infty e^{-sL} \frac{Q_N(L)}{N!}\, dL$$

$$= \frac{\exp[-(N\nu\gamma/2)]}{s^2} \int_{-\infty}^{\infty} \cdots \int_{-\infty}^{\infty} \exp\left(\nu\gamma \sum_1^N x_i\right) W(x_1)$$

$$\times \prod_{i=1}^{N-1} p_s(x_i|x_{i+1})\, dx_1, \ldots, dx_N \tag{7.10}$$

where

$$p_s(x|y) = \int_\delta^\infty e^{-st} P_\gamma(x|y;t)\, dt. \tag{7.11}$$

Introducing the kernel*

$$K_s(x,y) = \exp\left(\frac{\nu\gamma}{2} x\right) \frac{W(x) p_s(x|y)}{(W(x)W(y))^{1/2}} \exp\left(\frac{\nu\gamma}{2} y\right) \tag{7.12}$$

we can rewrite (7.10) in the form

$$\int_0^\infty e^{-sL} \frac{Q_N(L)}{N!}\, dL$$

$$= \frac{\exp[-(N\nu\gamma/2)]}{s^2} \int_{-\infty}^{\infty} \cdots \int_{-\infty}^{\infty} \exp\left(\frac{\nu\gamma}{2} x_1\right) W(x_1)$$

$$\times \prod_{i=1}^{N-1} K_s(x_i, x_{i+1}) \exp\left(\frac{\nu\gamma}{2} x_N\right) W(x_N)\, dx, \ldots, dx_N. \tag{7.13}$$

---

* The kernel $K_s(x, y)$ depends, of course, also on γ (as does $p_s(x \mid y)$) but I suppress this dependence to simplify the appearance of the formulas.

The kernel $K_s(x, y)$ can easily be shown to be symmetric, positive definite, and Hilbert-Schmidt. All standard theorems are applicable and, in particular, the right-hand side of (7.13) is easily expressible in terms of the eigenvalues (in decreasing order)

$$\lambda_1, \lambda_2, \ldots,$$

and the corresponding normalized eigenfunctions

$$\phi_1(x), \phi_2(x), \ldots,$$

of K_s.

In fact, we have

$$\int_0^\infty e^{-sL} \frac{Q_N(L)}{N!}\, dL = \frac{\exp[-(N\nu\gamma/2)]}{s^2} \sum_{j=1}^\infty \lambda_j^{N-1} \times \left(\int_{-\infty}^\infty \sqrt{W(x)} \exp\left(\frac{\nu\gamma}{2} x\right)\phi_j(x)\, dx\right)^2. \quad (7.14)$$

It follows at once that

$$\lim_{N\to\infty} \frac{1}{N} \log \int_0^\infty e^{-sL} \frac{Q_N(L)}{N!}\, dL = -\frac{\nu\gamma}{2} + \log \lambda_1(s)^* \quad (7.15)$$

and I shall now relate the limit on the left-hand side of (7.15) to thermodynamic quantities.

I shall do it formally though a rigorous proof is not too difficult to supply (see, e.g., Kac [5] where a proof based on the so-called grand canonical formalism can be found).

Set $L = Nl$ obtaining

$$\int_0^\infty e^{-sL} \frac{Q_N(L)}{N!}\, dL = N \int_0^\infty e^{-sNl} \frac{Q_N(Nl)}{N!}\, dl$$

---

* λ_1 depends not only on s but on γ and ν as well. We emphasize the dependence on s since (in this section at least) it matters most.

and recall that by the theorem of Ruelle

$$\lim_{N\to\infty} \frac{1}{N} \log \frac{Q_N(Nl)}{N!}$$

exists and can be identified with

$$-\beta f(l, \beta)$$

where f is the free energy per particle and l (the specific length) is clearly the one-dimensional counterpart of the specific volume v.

Thus

$$\frac{Q_N(Nl)}{N!} \sim \exp\{-N\beta f(l, \beta)\}$$

and it requires only a little optimism to conclude that

$$\int_0^\infty e^{-sL} \frac{Q_N(L)}{N!}\, dL \sim N \int_0^\infty \exp\{N(-sl - \beta f(l, \beta)\}\, dl.$$

Applying the familiar method of Laplace we are led to conclude that

$$\lim_{N\to\infty} \frac{1}{N} \log \int_0^\infty e^{-sL} \frac{Q_N(L)}{N!}\, dL = \operatorname*{Max}_{l} \{-sl - \beta f(l, \beta)\}. \quad (7.16)$$

Combining this with (7.15) we have

$$-\frac{\nu\gamma}{2} + \log \lambda_1(s) = \operatorname*{Max}_{l} \{-sl - \beta f(l, \beta)\}. \quad (7.17)$$

The l which maximizes $-sl - \beta f(l, \beta)$ can be found by solving the equation

$$-s - \beta \frac{\partial f}{\partial l} = 0 \quad (7.18)$$

and it is natural to define pressure by the analogue of (5.12), i.e.,

$$p = -\frac{\partial f}{\partial l}. \quad (7.19)$$

Equation (7.18) thus becomes

$$s = \beta p = \frac{p}{kT} \tag{7.20}$$

and equation (7.17) assumes the form

$$-\frac{\nu\gamma}{2} + \log \lambda_1(\beta p) = -\beta p l - \beta f(l, \beta). \tag{7.21}$$

Differentiating (7.21) with respect to l we obtain

$$l = -\frac{\lambda_1'(\beta p)}{\lambda_1(\beta p)} \tag{7.22}$$

which is the equation of state of our gas in an implicit form.*

It is a little more convenient to rewrite (7.22) in the form

$$l = -\frac{\lambda_1'(s)}{\lambda_1(s)}, \qquad s = \beta p = \frac{p}{kT}. \tag{7.23}$$

Note that s which up to now was only the variable in the Laplace transform has become intimately related to the pressure ($s = p/kT$). This is yet another confirmation of an observation of Schrödinger (made on p. 37 of his little book *Statistical Thermodynamics*) that "one of the fascinating features of statistical thermodynamics is that quantities and functions introduced primarily as mathematical devices almost invariably acquire a fundamental physical meaning."

8. It is easy to show that for $\gamma > 0$ the largest eigenvalue $\lambda_1(s)$ is a smooth function of s and ν so that the isotherms (whose equations are given by (7.22)) are also smooth and the model can therefore exhibit no condensation.

---

* In going from (7.21) to (7.22) we have assumed that dl/dp exists. Since p is a monotonic function of l this derivative exists for all p except perhaps for a set of measure zero. Thus it would seem that (7.22) has been justified only almost everywhere. However, since $\lambda_1(s)$ and $\lambda_1'(s)$ are continuous functions of s and since p is a continuous function of l (this too has been proved by Ruelle in great generality) formula (7.22) holds for *all* p (and all $l > \delta$).

This is, however, no longer so if we let γ approach zero.

In fact, we have

$$\lim_{\gamma \to 0} \lambda_1(s) = \omega(s) = \operatorname*{Max}_{-\infty<\eta<\infty} \left\{ \exp(\eta\sqrt{2\nu}) \frac{\exp(-\delta[s + (\eta^2/2)])}{s + (\eta^2/2)} \right\} \tag{8.1}$$

and, as we shall see, if

$$\nu > \frac{27\delta}{8}$$

the derivative $\omega'(s)$ of $\omega(s)$ has always exactly one discontinuity.

To prove (8.1) we note that $K_s(x, y)$ can be rewritten in the form

$$K_s(x, y) = \int_\delta^\infty \exp\left(-st + \frac{\gamma t}{2}\right) A(t, x) \frac{\exp\{-(y - x)^2/4 \sinh \gamma t\}}{\sqrt{4\pi \sinh \gamma t}} A(t, y)\, dt \tag{8.2}$$

where

$$A(t, x) = \exp\left\{ -\frac{1}{4} \left(\tanh \frac{\gamma t}{2}\right) x^2 + \frac{\sqrt{\nu\gamma}}{2} x \right\}.$$

Thus for $\phi \in L^2(-\infty, \infty)$ and $s > \gamma/2$ we have

$$\begin{aligned}
(\phi, K_s\phi) &= \int_{-\infty}^\infty \int_{-\infty}^\infty \phi(x) K_s(x, y) \phi(y)\, dx\, dy \\
&\leqslant \int_\delta^\infty \exp\left(-st + \frac{\gamma t}{2}\right) \int_{-\infty}^\infty \phi^2(x) \\
&\quad \times \exp\left\{ -\frac{1}{2} \left(\tanh \frac{\gamma t}{2}\right) x^2 + \sqrt{\nu\gamma} x \right\} dx\, dt \\
&\leqslant \left(\int_{-\infty}^\infty \phi^2(x)\, dx \right) \operatorname*{Max}_x \left\{ \int_\delta^\infty \exp\left(-st + \frac{\gamma t}{2}\right) \right. \\
&\qquad \left. \times \exp\left\{ -\frac{1}{2} \left(\tanh \frac{\gamma t}{2}\right) x^2 + \sqrt{\nu\gamma} x \right\} dt \right\}.
\end{aligned} \tag{8.3}$$

Setting

$$x = \frac{\eta\sqrt{2}}{\sqrt{\gamma}}$$

(8.3) becomes

$$(\phi, K_s\phi) \leqslant \operatorname*{Max}_{\eta}\left\{\int_{\delta}^{\infty} \exp\left(-st + \frac{\gamma t}{2}\right) \times \exp\left\{-\frac{\tanh(\gamma t/2)}{\gamma}\eta^2 + \sqrt{2\nu}\eta\right\} dt\right\}(\phi, \phi)$$

and consequently* we obtain that

$$\limsup_{\gamma\to 0} \lambda_1(s) \leqslant \operatorname*{Max}_{\eta}\left\{\exp(\eta\sqrt{2\nu})\,\frac{\exp(-\delta[s + (\eta^2/2)])}{s + (\eta^2/2)}\right\} = \omega(s). \tag{8.4}$$

If we now set

$$\phi(y) = g\left(y - \frac{\bar{\eta}\sqrt{2}}{\sqrt{\gamma}}\right) \tag{8.5}$$

where $\bar{\eta}$ is the value of η for which the maximum on the right-hand side of (8.4) is actually achieved and expand $(K_s\phi, \phi)$ in powers of γ we get (with some labor) that

$$(K_s\phi, \phi) = \omega(s)\{(g, g) - \gamma g(g, Lg) + \cdots\} \tag{8.6}$$

where

$$Lg = \int_{-\infty}^{\infty}\left\{(g'(x))^2 + \left[-\frac{1}{2} + \left(\frac{1}{4} - \frac{\nu(q-\delta)^2}{2q^3}\right)x^2\right]g^2(x)\right\}dx \tag{8.7}$$

and

$$q = \frac{\sqrt{2\nu}}{\bar{\eta}}. \tag{8.8}$$

---

* Since for small γ we have $\tanh(\gamma t/2) \sim \gamma t/2$.

We have $\lambda_1(s) \geqslant (K_s\phi, \phi)/(\phi, \phi)$ and since $(\phi, \phi) = (g, g)$ one obtains that

$$\liminf_{\gamma\to 0} \lambda_1(s) \geqslant \omega(s). \tag{8.9}$$

Combining this with (8.4) we finally obtain (8.1), i.e.,

$$\lim_{\gamma\to 0} \lambda_1(s) = \omega(s). \tag{8.10}$$

9. The discussion of $\omega(s)$

$$\omega(s) = \operatorname*{Max}_{\eta} \left\{\exp(\eta\sqrt{2\nu})\, \frac{\exp(-\delta[s + (\eta^2/2)])}{s + (\eta^2/2)}\right\} \tag{9.1}$$

involves nothing more than elementary calculus and can be found in Kac, Uhlenbeck and Hemmer [6].* Here I shall summarize the pertinent results, adding here and there a few more details.

(a) If

$$\nu < \frac{27\delta}{8}$$

the function

$$F_s(\eta) = \exp(\eta\sqrt{2\nu})\, \frac{\exp(-\delta[s + (\eta^2/2)])}{s + (\eta^2/2)} \tag{9.2}$$

has a *unique* maximum at $\bar{\eta}(s)$ so that

$$\omega(s) = F_s(\bar{\eta}(s)) \tag{9.3}$$

and as a function of s $\omega(s)$ is real analytic and decreasing.

Moreover

$$-\frac{\omega'(s)}{\omega(s)} = l, \tag{9.4}$$

---

* It goes without saying that it is sufficient to restrict oneself to $\eta > 0$.

which by (7.23) and (8.1) is the limiting ($\gamma \to 0$) equation of state, is easily seen to be equivalent to the famed van der Waals equation.

$$s = \frac{p}{kT} = \frac{1}{l - \delta} - \frac{\nu}{l^2}. \tag{9.5}$$

(b) If

$$\nu > \frac{27\delta}{8} \tag{9.6}$$

the situation is more complicated. The equation which determines the location of the maxima is

$$\sqrt{2\nu} = \eta\left(\delta + \frac{1}{s + (\eta^2/2)}\right) = h_s(\eta) \tag{9.7}$$

which, depending on s, can have either one or three positive roots (if $\nu < 27\delta/8$ (9.7) has only one real root).

In particular, there are three roots if and only if

$$8s\delta < 1 \qquad \text{and} \qquad \frac{(3 + \sqrt{1 - 8s\delta})^3}{1 + \sqrt{1 - 8s\delta}} < \frac{8\nu}{\delta} < \frac{(3 - \sqrt{1 - 8s\delta})^3}{1 - \sqrt{1 - 8s\delta}}. \tag{9.8}$$

On the other hand, for sufficiently small or sufficiently large s, and for a fixed ν satisfying (9.6), there is only one real (positive) root.

This situation is illustrated in Figure 5, and one sees that as s increases from 0 the abscissa of the maximum goes over into the smallest of the three roots $\bar{\eta}_1(s)$ while as s decreases from ∞ the abscissa of the maximum goes over into the largest one $\bar{\eta}_2(s)$. It is easily seen that $\bar{\eta}_1(s)$ is not defined if the maximum of $h_s(\eta)$ falls below $\sqrt{2\nu}$. This happens for

$$s > s_1(\nu) \tag{9.9}$$

where $s_1(\nu)$ is the (unique) root of the equation

$$\frac{8\nu}{\delta} = \frac{(3 - \sqrt{1 - 8s\delta})^3}{1 - \sqrt{1 - 8s\delta}}. \tag{9.10}$$

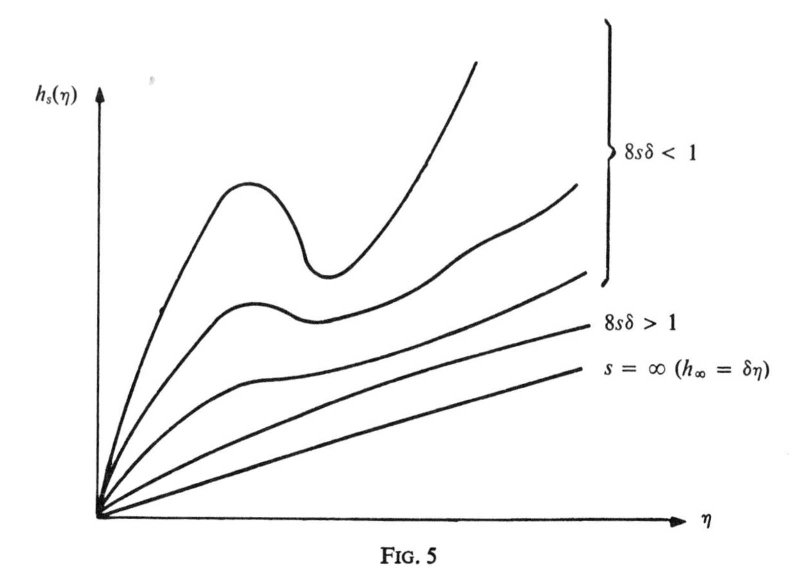

FIG. 5

On the other hand, $\bar{\eta}_2(s)$ is defined only whenever the minimum of $h_s(\eta)$ is lower than $\sqrt{2\nu}$, i.e., when

$$\frac{8\nu}{\delta} > \frac{(3 + \sqrt{1 - 8s\delta})^3}{1 + \sqrt{1 - 8s\delta}}. \tag{9.11}$$

Since the expression on the right-hand side of (9.11) is a decreasing function of s (or better yet of $s\delta$) we see that (9.11) is satisfied for all $s > 0$ if $8\nu/\delta > 32$ or if $27 \leqslant 8\nu/\delta \leqslant 32$ only for

$$s > s_2(\nu) \tag{9.12}$$

where $s_2(\nu)$ is the (unique) root of the equation

$$\frac{8\nu}{\delta} = \frac{(3 + \sqrt{1 - 8s\delta})^3}{1 + \sqrt{1 - 8s\delta}}. \tag{9.13}$$

It should be noted that

$$s_2(\nu) < s_1(\nu)$$

and that the graphs of $F_s(\bar{\eta}_1(s))$ and $F_s(\bar{\eta}_2(s))$ look as shown in Figure 6, with the understanding that $s_2(\nu)$ should be replaced by 0 if $8\nu/\delta > 32$.

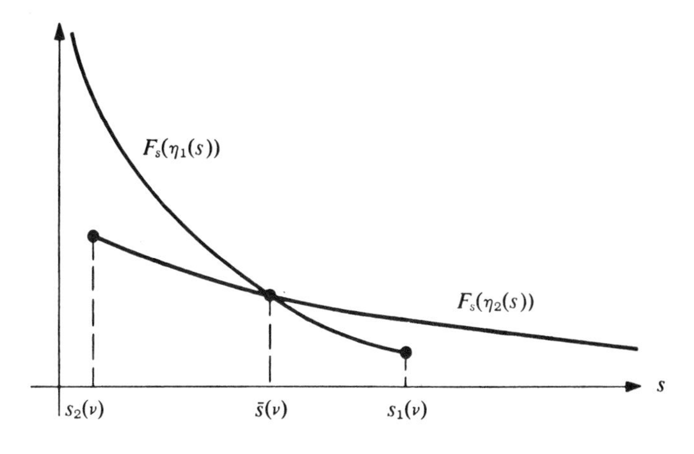

FIG. 6

The fact that the two graphs intersect requires justification which can be provided as follows.

For $s_2(\nu) < s < s_1(\nu)$ both η_1 and η_2 are defined and we have, of course,

$$\frac{\sqrt{2\nu}}{\bar{\eta}_1(s)} = \delta + \frac{1}{s + \bar{\eta}_1^2(s)/2} \tag{9.14}$$

and

$$\frac{\sqrt{2\nu}}{\bar{\eta}_2(s)} = \delta + \frac{1}{s + \bar{\eta}_2^2(s)/2}. \tag{9.15}$$

What we have to show is that the equation

$$F(\bar{\eta}_1(s)) = F(\bar{\eta}_2(s)) \tag{9.16}$$

has a solution for some $\bar{s} = \bar{s}(\nu)$. Set

$$l_1 = \frac{\sqrt{2\nu}}{\bar{\eta}_2(s)}, \qquad l_2 = \frac{\sqrt{2\nu}}{\bar{\eta}_1(s)}* \tag{9.17}$$

* Since $\bar{\eta}_1 < \bar{\eta}_2$ one has $l_2 > l_1$ in analogy with Fig. 4.

so that (9.14) and (9.15) become

$$s = \frac{1}{l_2 - \delta} - \frac{\nu}{l_2^2} \tag{9.14a}$$

and

$$s = \frac{1}{l_1 - \delta} - \frac{\nu}{l_1^2} \tag{9.15a}$$

respectively.

Equation (9.16) assumes now, upon taking logarithms of both sides, the form

$$\log(l_1 - \delta) + \frac{2\nu}{l_1} - \frac{\delta}{l_1 - \delta} = \log(l_2 - \delta) + \frac{2\nu}{l_2} - \frac{\delta}{l_2 - \delta}. \tag{9.18}$$

Using (9.14a) and (9.14b) we verify that

$$\frac{2\nu}{l_1} - \frac{\delta}{l_1 - \delta} = 1 - l_1 s + \frac{\nu}{l_1} \tag{9.19}$$

and

$$\frac{2\nu}{l_2} - \frac{\delta}{l_2 - \delta} = 1 - l_2 s + \frac{\nu}{l_2} \tag{9.20}$$

which upon substituting into (9.18) yields

$$(l_1 - l_2)s = \int_{l_1}^{l_2} \left(\frac{1}{l - \delta} - \frac{\nu}{l^2}\right) dl. \tag{9.21}$$

It remains only to show that s as defined by (9.21) falls within the range between $s_2(\nu)$ and $s_1(\nu)$.

If we plot the graph of

$$s = \frac{1}{l - \delta} - \frac{\nu}{l^2}, \qquad l > \delta, \tag{9.22}$$

we note that for $\nu > 27\delta/8$ it has the appearance shown in Figure 7, and (9.21) defines that s (i.e., $\bar{s}$) for which the shaded areas are equal. The reader may now check that $s_1(\nu)$ is the height of the maximum and $s_2(\nu)$ the height of the minimum of the curve (9.22) and the inequality $s_2(\nu) < \bar{s} < s_1(\nu)$ becomes trivial.

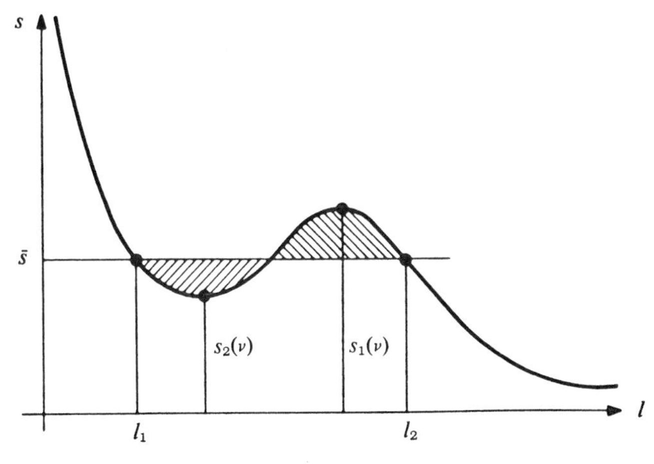

FIG. 7

Summarizing the discussion we can say that for $\nu > 27\delta/8$ the graph of $\omega(s) = \text{Max}(F(\bar{\eta}_1(s), F(\bar{\eta}_2(s))$ consists of two smooth pieces which join continuously at $\bar{s} = \bar{s}(\nu)$ with a discontinuity in the slope (see Figure 8). The limiting equation of state

$$-\frac{\omega'(s)}{\omega(s)} = l$$

owing to the discontinuity in the slope can be represented graphically as in Figure 9, or, turning the graph around, we obtain a graph just as in Figure 4. For $l < l_1$ and $l > l_2$ the equation is still the van der Waals equation

$$s = \frac{p}{kT} = \frac{1}{l - \delta} - \frac{\nu}{l^2} \tag{9.23}$$

but for $l_1 < l < l_2$ the "wiggle" implied by (9.23) must be replaced by a horizontal line drawn at the height $\bar{s}$ determined by the "equal area rule" of Maxwell (9.21).

10. It will be instructive to compare the derivation given above with a derivation, due to L. S. Ornstein [7], and which in the most

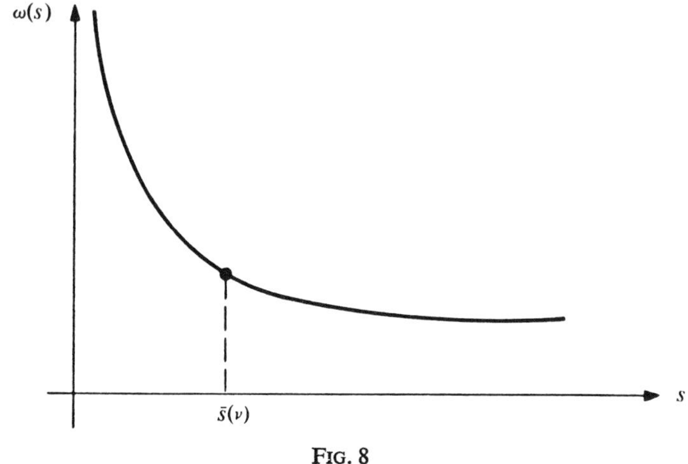

FIG. 8

direct and explicit way embodies the basic assumption of the van der Waals theory that attractive forces are weak and long range.

In the context of our model (though it has much wider applicability) the idea is to replace in the exact partition function (7.2), i.e.,

$$\frac{Q_N(L)}{N!} = \frac{1}{N!}\int_0^L \cdots \int_0^L \prod_{1 \le i < j \le N} S_\delta(|t_j - t_i|) \times \exp\left(\nu\gamma \sum_{1 \le i < j \le N} e^{-\gamma|t_j - t_i|}\right) d\Omega, \tag{10.1}$$

each term

$$\gamma e^{-\gamma|t_j - t_i|} \tag{10.2}$$

by an appropriate constant c. To find this constant, one argues as follows: In "general" t_i will be "well inside" the container (i.e., the interval $(0, L)$ so that the graph of (10.2) as a function of t_j looks as shown in Figure 10, and the constant c is determined by requiring that the shaded area under c be equal to the actual area under (10.2), i.e.,

$$cL = \gamma \int_0^L \exp\{-\gamma|t_j - t_i|\}\, dt_j$$

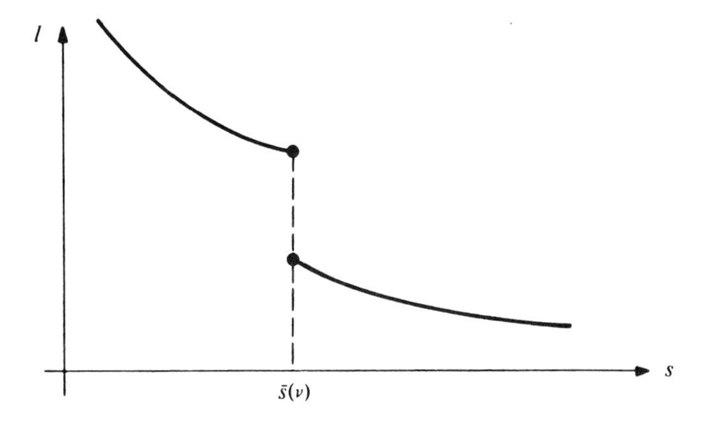

FIG. 9

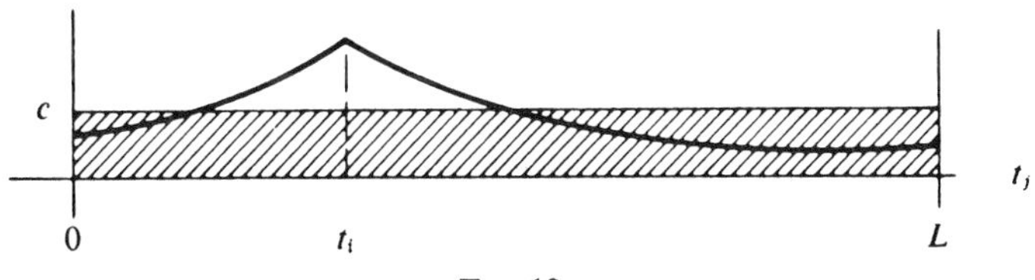

FIG. 10

or, since t_i is "well inside" $(0, L)$ and L is very large, we have

$$cL \sim \gamma \int_{-\infty}^{\infty} \exp(-\gamma|t|)\, dt = 2.$$

In other words, we can take

$$c = \frac{2}{L} \tag{10.3}$$

and replace (10.1) by

$$\frac{\tilde{Q}_N(L)}{N!} = \exp\left\{\frac{\nu}{L} N(N-1)\right\} \frac{1}{N!} \times \int_0^L \cdots \int_0^L \prod_{1 \le i < j \le N} S_\delta(|t_j - t_i|)\, d\Omega. \tag{10.4}$$

It should be clear that replacing $\gamma \exp\{-\gamma|t_j - t_i|\}$ by a constant can be reasonable *only* if γ (or rather the dimensionless number $\gamma\delta$) *is small* for only then is the graph of the interparticle potential flat. Small $\gamma\delta$ means, of course, that the attractive force is weak and long-range, and this was the fundamental assumption of van der Waals.

One recognizes that

$$\frac{1}{N!}\int_0^L \cdots \int_0^L \prod_{1 \le i < j \le N} S_\delta(|t_j - t_i|)\, d\Omega \tag{10.5}$$

is the partition function of the "gas of hard rods" which is a special case ($\alpha = 0$, $\nu = 0$) of our gas.

Formula (7.17) is still applicable but the integral equation is now elementary and one has

$$\lambda_1(s) = \frac{e^{-s\delta}}{s}. \tag{10.5}$$

It is also easy to show by a direct calculation that (10.5) is equal to $(L - N\delta)^N$.

In the thermodynamic limit

$$N \to \infty, \qquad L \to \infty, \qquad L/N = l$$

one then obtains quite easily the equation of state

$$\frac{p}{kT} = \frac{1}{l - \delta} - \frac{\nu}{l^2} \tag{10.6}$$

which is the van der Waals equation.

As we have seen above for $\nu > 27\delta/8$ the equation (10.6) leads to a "wiggle" or a loop and this contradicts the Second Law of Thermodynamics which requires that pressure be a nonincreasing function of l (see (2.4)).

To resolve this difficulty Maxwell modified the van der Waals equation by replacing the "wiggle" by the horizontal piece. To determine the height at which to place the horizontal portion Maxwell used a *purely thermodynamic* argument and was led to the "equal area" rule discussed a while back.

Why dwell on an argument which would repel a mathematician by its crudity and "uncleanliness"?

To answer this question let me first quote from Maxwell's review of van der Waals' Thesis:

> The molecular theory of continuity of the liquid and gaseous states forms the subject of an exceedingly ingenious dissertation by Mr. Johannes Diderick van der Waals, a graduate of Leyden. There are certain points in which I think he has fallen into mathematical errors, and his final result is certainly not a complete expression for the interaction of real molecules, but his attack on this difficult question is so able and so brave, that it cannot fail to give a notable impulse to molecular science. It has certainly directed the attention of more than one inquirer to the study of the Low-Dutch language in which it is written.

This quote serves to emphasize the oft forgotten fact that what may appear as mathematically crude and logically unclean can also be an "able and brave attack" on a difficult question posed by nature.

But this is not the whole answer. Take a look at the kernel $K_s(x, y)$ in formula (7.12).

As $\gamma \to 0$ it approaches the kernel

$$\frac{W(x)p_s(x|y)}{(W(x)W(y))^{1/2}}, \tag{10.7}$$

and one would be inclined to think that the maximum eigenvalue $\lambda_1(s)$ of (7.12) approaches the maximum eigenvalue of (10.7) which happens to be $\exp(-s\delta)/s$. We know, of course, from the analyses of Sections 7 and 8 that this is not the case and that moreover the analytic form of the limit (as $\gamma \to 0$) of $\lambda_1(s)$ undergoes a striking change at the critical value $\nu = 27\delta/8$. Could one have suspected this a priori? Perhaps. But a "physical" derivation like that of Ornstein is certainly highly suggestive and is most helpful in guiding us toward a proper mathematical formulation.

Actually one sees that in the Ornstein derivation the thermodynamic limit ($N \to \infty$, $L \to \infty$, $L/N = l$) is *coupled* with the limit $\gamma \to 0$ (the van der Waals limit). The price for this transgression against the proper way of taking limits is the "wiggle" and the resulting violation of the Second Law. An even greater price is paid if one first takes the limit $\gamma \to 0$ and then the thermodynamic limit.

For with this order of taking the limits one obtains simply the gas of hard rods, and the whole phenomenon of condensation is lost.

In Statistical Mechanics the order in which various limits are taken is of utmost importance, and this fact alone should appeal to an analyst.

11. Let us recall (see [8.6] and [8.7]) that

$$(K_s\phi, \phi) = \omega(s)\{(g, g) - \gamma g(Lg, g) + \cdots\} \tag{11.1}$$

where

$$\phi(x) = g\left(x - \frac{\bar{\eta}\sqrt{2}}{\sqrt{\gamma}}\right) \tag{11.2}$$

and

$$(Lg, g) = \int_{-\infty}^{\infty} \left\{g'^2(x) + \left[-\frac{1}{2} + \left(\frac{1}{4} - \frac{\nu(q - \delta)}{2q^3}\right)^2 x^2\right] g^2(x)\right\} dx. \tag{11.3}$$

It should also be recalled that

$$q = \frac{\sqrt{2\nu}}{\bar{\eta}}. \tag{11.4}$$

If the function g is chosen so as to minimize (11.3) under the constraint $(g, g) = 1$, i.e., if we take

$$g(x) = C \exp\left\{-\frac{1}{4}\sqrt{1 - \frac{2\nu(q - \delta)^2}{q^3}}\, x^2\right\}, \tag{11.5}$$

with C the normalizing constant, then the function

$$C \exp\left\{-\frac{1}{4}\sqrt{1 - \frac{2\nu(q - \delta)^2}{q^3}} \left(x - \frac{\bar{\eta}\sqrt{2}}{\sqrt{\gamma}}\right)^2\right\} \tag{11.6}$$

is an approximate principal eigenfunction of K_s, accurate enough to yield $\lambda_1(s)$ *correct to order* γ.

However, for $\nu > 27\,\delta/8$ this is true only if $s \neq \bar{s}(\nu)$. If $s = \bar{s}(\nu)$ there are two distinct η's; namely, $\bar{\eta}_1(\bar{s})$ and $\bar{\eta}_2(\bar{s})$ for which (see (9.2))

$$\omega(\bar{s}) = F_{\bar{s}}(\bar{\eta}_1(\bar{s})) = F_{\bar{s}}(\bar{\eta}_2(\bar{s})) \tag{11.7}$$

and consequently both

$$\phi_1^{(1)} = C_1 \exp\left\{-\frac{1}{4}\sqrt{1 - \frac{2\nu(q_1 - \delta)^2}{q_1^3}}\left(x - \frac{\bar{\eta}_1\sqrt{2}}{\sqrt{\gamma}}\right)^2\right\} \tag{11.8}$$

and

$$\phi_1^{(2)} = C_2 \exp\left\{-\frac{1}{4}\sqrt{1 - \frac{2\nu(q_2 - \delta)^2}{q_2^3}}\left(x - \frac{\bar{\eta}_2\sqrt{2}}{\sqrt{\gamma}}\right)^2\right\} \tag{11.9}$$

as well as every linear combination

$$\alpha_1\phi_1^{(1)} + \alpha_2\phi_1^{(2)}, \ \alpha_1^2 + \alpha_2^2 = 1 \tag{11.10}$$

are approximate eigenfunctions which yield the *same* approximations to λ_1 (i.e., the one which is correct to order γ).

The fact that *to order* γ

$$(K_s\phi_1^{(1)}, \phi_1^{(1)}) = (K_s\phi_1^{(2)}, \phi_1^{(2)})$$

is obvious as well as the facts that (again to order γ)

$$K_s\phi_1^{(1)} = \lambda_1\phi_1^{(1)}, \qquad K_s\phi_1^{(2)} = \lambda_1\phi_1^{(2)}.$$

Since, as can be easily seen, $(\phi_1^{(1)}, \phi_1^{(2)})$ is of the order $\exp(-\text{const}/\gamma)$, we can say that $\phi_1^{(1)}$ and $\phi_1^{(2)}$ are orthogonal to *all orders in* γ; and in particular, it follows that $\alpha_1\phi_1 + \alpha_2\phi_2$, $\alpha_1^2 + \alpha_2^2 = 1$ is also an approximate eigenfunction.

We are dealing here with an instant of an interesting phenomenon known as *asymptotic degeneracy*. As we shall see in the next section, it is this degeneracy that is directly connected with the question of geometric separation of coexisting phases.

12. As we have shown in Section 6 one can make geometric separation of phases plausible by proving the linear combination property (6.9) of distribution functions.

For our one-dimensional model everything can be exhaustively and rigorously discussed.

First of all, one can derive formulas for Laplace transforms of distribution functions.*

In the thermodynamic limit ($N \to \infty$, $L \to \infty$, $L/N = l$) the distribution functions $\bar{n}_k(x_1, x_2, \ldots, x_k; l)$ depend only on the mutual

---

* For details, see Uhlenbeck, Hemmer, and Kac [8].

distances $|x_i - x_j|$ and if $0 < x_1 < x_2 < \cdots < x_k$ one actually has

$$\bar{n}_k(x_1\, x_2, \ldots, x_k; l) \equiv \bar{\bar{n}}_k(x_2 - x_1, x_3 - x_2, \ldots, x_k - x_{k-1}; l). \quad (12.1)$$

What one can do is to calculate the multiple Laplace transforms

$$\int_0^\infty \cdots \int_0^\infty \exp\left(-\sum_1^{k-1} \sigma_j \zeta_j\right) \bar{\bar{n}}_k(\zeta_1, \zeta_2, \ldots, \zeta_{k-1}; l)\, d\zeta_1, \ldots, d\zeta_{k-1} \quad (12.2)$$

in terms of the resolvent of kernel K_s.

Set

$$R_s(x, y, \zeta) = \sum_{m=0}^{\infty} \zeta^m K_s^{(m)} \quad (12.3)$$

where $K_s^{(0)} \equiv K_s$ and $K_s^{(r)}$ is rth iterate of K_s. In other words,

$$R_s = (I - \zeta K_s)^{-1}. \quad (12.4)$$

We then have

$$l \int_0^\infty \exp(-\sigma\zeta)\bar{\bar{n}}_2(\zeta; l)\, d\zeta$$

$$= \frac{1}{\lambda_1(s)} \int_{-\infty}^\infty \int_{-\infty}^\infty \phi_1(x) R_{s+\sigma}\left(x, y, \frac{1}{\lambda_1(s)}\right) \phi_1(y)\, dx\, dy \quad (12.5)$$

$$l \int_0^\infty \int_0^\infty \exp(-\sigma_1\zeta_1 - \sigma_2\zeta_2)\bar{\bar{n}}_3(\zeta_1, \zeta_2; l)\, d\zeta_1\, d\zeta_2$$

$$= \frac{1}{\lambda_1^2(s)} \int_{-\infty}^\infty \int_{-\infty}^\infty \int_{-\infty}^\infty \phi_1(x_1) R_{s+\sigma_1}\left(x_1, x_2, \frac{1}{\lambda_1(s)}\right)$$

$$\times R_{s+\sigma_2}\left(x_2, x_3, \frac{1}{\lambda_1(s)}\right) \phi_1(x_3)\, dx_1\, dx_2\, dx_3 \quad (12.6)$$

and so on for higher distribution functions.

In these formulas ϕ_1 is the principal eigenfunction of K_s, i.e., the eigenfunction corresponding to the largest eigenvalue λ_1.

Suppose now that*

$$l = \xi_1 l_1 + \xi_2 l_2, \qquad \xi_1 + \xi_2 = 1, \qquad \xi_1 \geqslant 0, \qquad \xi_2 \geqslant 0. \quad (12.7)$$

(See Figure 11.)

---

* It should be noted that $l_1 = q_1$ and $l_2 = q_2$.

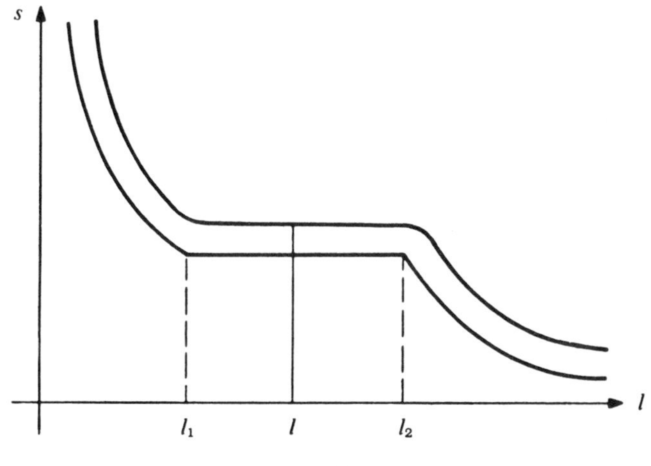

FIG. 11

The exact equation of state

$$\lambda_1'(s) = -l\lambda_1(s)$$

can be written in the form

$$-l\lambda_1(s) = \int_{-\infty}^{\infty}\int_{-\infty}^{\infty} \phi_1(x)\,\frac{\partial K_s}{\partial s}\,\phi_1(y)\,dx\,dy. \tag{12.8}$$

This follows by noting that

$$\lambda_1(s) = \int_{-\infty}^{\infty}\int_{-\infty}^{\infty} \phi_1(x)K_s\phi_1(y)\,dx\,dy$$

and that

$$\int_{-\infty}^{\infty}\int_{-\infty}^{\infty} \frac{\partial\phi_1}{\partial s}\,K_s\phi_1(y)\,dx\,dy$$

$$= \lambda_1(s)\int_{-\infty}^{\infty} \frac{\partial\phi_1(x)}{\partial s}\,\phi_1(x)\,dx = \frac{1}{2}\,\lambda_1\,\frac{\partial}{\partial s}\int_{-\infty}^{\infty} \phi_1^2(x)\,dx$$

$$= 0.$$

As $\gamma \to 0$ $s(l)$ approaches $\bar{s}(\nu)$ and $\lambda_1(s)$ approaches $\omega(\bar{s})$. What about ϕ_1?

If one notes that

$$\lim_{\gamma \to 0} \int_{-\infty}^{\infty} \int_{-\infty}^{\infty} \phi_1^{(1)}(x) \frac{\partial K_s}{\partial s} \phi_1^{(1)}(y)\, dx\, dy = -l_1 \omega(\bar{s}) \tag{12.9}$$

and that

$$\lim_{\gamma \to 0} \int_{-\infty}^{\infty} \int_{-\infty}^{\infty} \phi_1^{(2)}(x) \frac{\partial K_s}{\partial s} \phi_1^{(2)}(y)\, dx\, dy = -l_2 \omega(\bar{s}), \tag{12.10}$$

we see that in order to remain consistent with (12.8) in the limit $\gamma \to 0$ we must take

$$\phi_1 = \alpha_1 \phi_1^{(1)} + \alpha_2 \phi_2^{(2)} \tag{12.11}$$

with

$$\alpha_1 = \sqrt{\xi_1}, \qquad \alpha_2 = \sqrt{\xi_2}. \tag{12.12}$$

If we now substitute (12.11) into (12.5), we obtain

$$\begin{aligned}
&l \int_0^{\infty} \exp(-\sigma\zeta) \bar{\bar{n}}_2(\zeta; l)\, d\zeta \\
&\quad = \frac{1}{\lambda_1(\bar{s})} \int_{-\infty}^{\infty} \int_{-\infty}^{\infty} (\sqrt{\xi_1}\phi_1^{(1)}(x) + \sqrt{\xi_2}\phi_1^{(2)}(x) \\
&\qquad \times R_{s+\sigma}\left(x, y; \frac{1}{\lambda_1(\bar{s})}\right)(\sqrt{\xi_1}\phi_1^{(1)}(y) + \sqrt{\xi_2}\phi_1^{(2)}(y))\, dx\, dy \\
&\quad = \xi_1 \frac{1}{\lambda_1(\bar{s})} \int_{-\infty}^{\infty} \int_{-\infty}^{\infty} \phi_1^{(1)}(x) R_{\bar{s}+\sigma}\left(x, y; \frac{1}{\lambda_1(\bar{s})}\right) \phi_1^{(1)}(y)\, dx\, dy \\
&\qquad + \xi_2 \frac{1}{\lambda_1(\bar{s})} \int_{-\infty}^{\infty} \int_{-\infty}^{\infty} \phi_1^{(2)}(x) R_{\bar{s}+\sigma}\left(x, y; \frac{1}{\lambda_1(\bar{s})}\right) \phi_1^{(2)}(y)\, dx\, dy \\
&\qquad + 2\sqrt{\xi_1 \xi_2} \frac{1}{\lambda_1(\bar{s})} \int_{-\infty}^{\infty} \int_{-\infty}^{\infty} \phi_1^{(1)}(x) R_{\bar{s}+\sigma}\left(x, y; \frac{1}{\lambda_1(\bar{s})}\right) \phi_1^{(2)}(y)\, dx\, dy.
\end{aligned} \tag{12.13}$$

It is now easy to show that as $\gamma \to 0$ we obtain

$$\lim_{\gamma\to 0} l \int_0^\infty e^{-\sigma\zeta}\bar{\bar{n}}_2(\zeta; l)\, d\zeta$$

$$= \xi_1 l_1 \int_{-\infty}^{\infty} e^{-\sigma\zeta}\bar{\bar{n}}_2^{(0)}(\zeta; l_1)\, d\zeta + \xi_2 l_2 \int_0^\infty e^{-\sigma\zeta}\bar{\bar{n}}_2^{(0)}(\zeta, l_2)\, d\zeta \quad (12.14)$$

where $\bar{\bar{n}}_2^{(0)}(\zeta; l)$ is the n_2 function corresponding to a gas of hard rods (with *no* attraction) of length δ whose Laplace transform is given by the formula

$$l \int_0^\infty e^{-\sigma\zeta}\bar{\bar{n}}_2^{(0)}(\zeta; l)\, d\zeta = \frac{1}{[1 + \sigma(l - \delta)]e^{\sigma\delta} - 1}. \quad (12.15)$$

Similarly $\bar{\bar{n}}_k$ for $k > 2$ become in the limit $\gamma \to 0$ line as combinations of corresponding functions for a gas of hard rods (at densities $1/l_1$ and $1/l_2$).

We therefore see that in the limit $\gamma \to 0$ we have indeed two coexisting (and geometrically separated) phases.

It should perhaps be mentioned that the Virial Theorem (see (6.10)) for our gas assumes the form

$$p = \frac{kT}{l} + \delta kT\bar{\bar{n}}_2(\delta+; l) - \alpha\gamma^2 \int_\delta^\infty d\zeta e^{-\gamma\zeta}\zeta\bar{\bar{n}}_2(\zeta; l) \quad (12.16)$$

and that it can be derived directly from our integral equation.

2. LATTICE GASES AND THE SPHERICAL MODELS

1. Before discussing lattice gases let me review briefly a formalism for deriving the equation of state based on the concept of the so-called grand canonical ensemble.

I shall restrict myself to the one-dimensional case, although it is not essential.

Let $Q_N(L)$ as usual denote the ordinary partition function and set

$$G(L; z) = \sum_{N=0}^{\infty} \frac{Q_N(L)}{N!} z^N \cdot (Q_0(L) = 1). \quad (2.1.1)$$

$G(L; z)$ is called the grand partition function.

Let us assume that

$$\frac{Q_N(L)}{N!} \sim e^{-N\beta f(l;\beta)} \tag{2.1.2}$$

in the thermodynamic limit (i.e., $N \to \infty$, $L \to \infty$, $L/N = l$), and consider the Laplace transform of G with respect to L, i.e.,

$$\int_0^\infty e^{-sL} G(L; z)\, dL = \sum_{N=0}^{\infty} z^N \int_0^\infty e^{-sL} \frac{Q_N(L)}{N!}\, dL. \tag{2.1.3}$$

As we have seen several times

$$\frac{1}{N} \log \int_0^\infty e^{-sL} \frac{Q_N(L)}{N!}\, dL \sim \operatorname*{Max}_l (-sl - \beta f(l; \beta)),$$

and therefore the series converges only for those real positive z which satisfy the inequality

$$z e^{\operatorname{Max}_l(-sl - \beta f(l;\beta))} \leqslant 1. \tag{2.1.4}$$

We can also say that for a *fixed* (real positive) z the above inequality determines the s for which the Laplace transform converges.

In other words the abscissa of convergence is determined by the equation

$$\operatorname*{Max}_l (-sl - \beta f(l; \beta)) = -\log z. \tag{2.1.5}$$

The $l = l(s)$ for which Max is attained is obtained from the equation

$$s = -\beta \frac{\partial f}{\partial l}, \tag{2.1.6}$$

and we have

$$-sl(s) - \beta f(l(s); \beta) = -\log z, \tag{2.1.7}$$

which determines s as a function of z. Differentiating with respect to z and taking into account that $s = -\beta\, \partial f/\partial l$, we obtain

$$\frac{1}{l} = z \frac{ds}{dz}. \tag{2.1.8}$$

It is, of course, understood that $s = \chi(z)$ is the abscissa of convergence of the Laplace transform.

Summarizing we have

$$s = \frac{p}{kT} = \chi(z)$$

$$\text{(G)}$$

$$\frac{1}{l} = z\chi'(z)$$

where $\chi(z)$ is the abscissa of convergence.

On the other hand by a classical result

$$\chi(z) = \lim_{L\to\infty} \frac{1}{L} \log G(L; z) \tag{2.1.9}$$

provided the limit exists (in general one must take the limit superior) and by eliminating z from Equation (G) above, one again obtains the equation of state.

2. A (one-dimensional) lattice gas is a collection of N points which can be placed at $1, 2, 3, \ldots, L$ $(N \leqslant L)$ and no two points can occupy the same place (hard-core assumption). In addition one has an attractive two-body interaction between the points so that the interaction energy is of the form

$$E = \sum_{1\le i<j\le N} \phi(|k_i - k_j|) \tag{2.2.1}$$

where $\phi(0) = +\infty$ (hard-core assumption), and

$$\phi(l) = -\rho(l), \qquad l = \pm 1, \pm 2, \ldots,$$

$$\rho \geqslant 0 \qquad \text{and} \qquad \rho(l) = \rho(-l).$$

One can now consider the partition function

$$Q_N(L) = \sum_{k_1=1}^{L} \cdots \sum_{k_N=1}^{L} e^{-\beta E} \tag{2.2.2}$$

and by analogy with the continuous case construct an equation of state. Let me do it via the grand canonical formalism and define

$$G(L; z) = \sum_{N=0}^{\infty} \frac{z^N}{N!} Q_N(L). \tag{2.2.3}$$

Instead of describing the state of the gas by actually specifying the positions $k_1, k_2, \ldots, k_N$ of the first, second, etc., particles, we can proceed a little differently and prescribe first the N (out of L) places to be occupied. This can be done by assigning to each place j a symbol ϵ_j,

$$\epsilon_j = \begin{cases} 0 \\ 1, \end{cases}$$

1 if j is occupied and 0 if it is not. The fact that there are N particles means that

$$\sum_{j=1}^{L} \epsilon_j = N$$

and the energy corresponding to a given set of ϵ's is clearly

$$E = -\sum_{1 \leq i < j \leq L} \rho(j-i)\epsilon_i\epsilon_j. \tag{2.2.4}$$

Since having prescribed N places there are $N!$ ways of arranging particles in them; we have

$$\frac{Q_N(L)}{N!} = \sum_{\epsilon_1+\epsilon_2+\cdots+\epsilon_L=N} \exp\left\{\beta \sum_{1 \leq i < j \leq L} \rho(j-i)\epsilon_i\epsilon_j\right\} \tag{2.2.5}$$

and

$$G(L; z) = \sum_{(\epsilon)} \exp\left\{\beta \sum_{1 \leq i < j \leq L} \rho(i-j)\epsilon_i\epsilon_j\right\} z^{\epsilon_1+\cdots+\epsilon_L} \tag{2.2.6}$$

where the summation is over *all* $\epsilon_1, \epsilon_2, \ldots, \epsilon_L$.

Set

$$\epsilon_j = \frac{\mu_j + 1}{2}, \qquad \mu_j = \begin{cases} +1 \\ -1 \end{cases}$$

and note that

$$\begin{aligned} \sum_{1 \leq i < j \leq L} \rho(i-j)\epsilon_i\epsilon_j &= \frac{1}{2}\sum_{i,j=1}^{L} \rho(i-j)\epsilon_i\epsilon_j - \frac{\rho(0)}{2}\sum_{1}^{L} \epsilon_j \\ &= \frac{1}{8}\sum_{i,j=1}^{L} \rho(i-j)\mu_i\mu_j + \frac{2}{8}\sum_{i,j=1}^{L} \rho(i-j)\mu_j \\ &\quad + \frac{1}{8}\sum_{i,j=1}^{L} \rho(i-j) - \frac{1}{4}\sum_{1}^{L} \mu_j - \frac{L}{4}. \end{aligned}$$

Although

$$\sum_{i=1}^{L} \rho(i-j)$$

depends on j, it is clear that the dependence for most j is weak and one can use the approximation

$$\sum_{i=1}^{L} \rho(i-j) \sim \sum_{-\infty}^{\infty} \rho(k) = R$$

while

$$\sum_{i,j=1}^{L} \rho(i-j) \sim LR.$$

Finally,

$$\begin{aligned}\chi(z) &= \lim_{L\to\infty} \frac{1}{L} \log G(L; z)\\ &= \lim_{L\to\infty} \frac{1}{L} \log \sum_{\vec{\mu}} \exp\left\{\frac{\beta}{8} \sum_{i,j=1}^{L} \rho(i-j)\mu_i\mu_j\right.\\ &\quad \left. + \left(\frac{R-\rho(0)}{4}\beta + \frac{1}{2}\log z\right) \sum_{k=1}^{L} \mu_k + \frac{R-2}{8}\beta + \frac{1}{2}\log z\right\}.\end{aligned} \tag{2.2.7}$$

Set

$$h = \frac{(R-\rho(0))}{4}\beta + \tfrac{1}{2}\log z. \tag{2.2.8}$$

The existence of the limit

$$f(h; \beta) = \lim_{L\to\infty} \frac{1}{L} \log \sum_{\vec{\mu}} \exp\left\{\frac{\beta}{8} \sum_{i,j=1}^{L} \rho(i-j)\mu_i\mu_j + h \sum_{k=1}^{L} \mu_k\right\} \tag{2.2.9}$$

under mild conditions on ρ has been proved by Lee and Yang [9].

If the equation of state of the lattice gas is to have a "flat part" there must be a discontinuity of $\chi'(z)$ for some z. In other words

$$\frac{\partial f}{\partial h},$$

must have a discontinuity for some h.

Since clearly

$$f(h; \beta) = f(-h; \beta)$$

one might expect that a discontinuity of $\partial f/\partial h$ will occur (if at all) at $h = 0$. Indeed again Lee and Yang have shown that $\partial f/\partial h$ is continuous for $h \neq 0$ and the question remains is it possible for $\partial f/\partial h$ to have a jump discontinuity at $h = 0$. The question has been answered in the affirmative by Dyson [10] (for ρ's satisfying an additional condition) but I shall not go into it here.

Instead I shall consider the problem for an artificial but highly suggestive modification of the lattice gas model.

The modification consists in replacing

$$\sum_{\vec{\mu}}$$

in the definition of F by

$$\int_{\Sigma\mu_j^2 = L} \cdots \, d\sigma,$$

i.e., allowing the μ's to vary continuously subject to the "spherical constraint" $\Sigma\mu_j^2 = L$ and with *uniform* density on the sphere in question.

I shall thus consider

$$\tilde{f}(h;\beta) = \lim_{L\to\infty} \frac{1}{L} \log \int_{\Sigma\mu_j^2 = L} \exp\left\{\frac{\beta}{8} \sum_{i,j=1}^{L} \rho(i-j)\mu_i\mu_j + h \sum_{j=1}^{L} \mu_j\right\} d\sigma \tag{2.2.10}$$

and show that under a certain additional condition on ρ

$$\frac{\partial \tilde{f}}{\partial h}$$

has a discontinuity at $h = 0$ for sufficiently large β. To calculate the modified partition function

$$\tilde{F}(h;\beta) = \int_{\Sigma\mu_j^2 = L} \exp\left\{\frac{\beta}{8} \sum_{i,j=1}^{L} \rho(i-j)\mu_i\mu_j + h \sum_{j=1}^{L} \mu_j\right\} d\sigma, \tag{2.2.11}$$

consider the integral

$$\int_{-\infty}^{\infty} \cdots \int_{-\infty}^{\infty} \exp\left(-s \sum_{j=1}^{L} x_j^2\right)$$

$$\times \exp\left\{\frac{\beta}{8} \sum_{i,j=1}^{L} \rho(i-j)x_i x_j + h \sum_{j=1}^{L} x_j\right\} dx_1, \ldots, dx_L \tag{2.2.12}$$

for sufficiently large positive s.

First note that the above integral is equal to

$$\int_0^\infty dr \int_{\sum_{j=1}^L x_j^2 = r^2} \exp\left(-s \sum_{j=1}^L x_j^2 + \frac{\beta}{8} \sum_{i,j=1}^L x_i x_j + h \sum_{j=1}^L x_j\right) d\sigma_r$$

$$= \int_0^\infty dr \exp(-sr^2) \int_{\sum x_j^2 = r^2} \exp\left\{\frac{\beta}{8} \sum_{i,j=1}^L \rho(i-j) x_i x_j + h \sum_{j=1}^L x_j\right\} d\sigma_r$$

$$= \sqrt{L} \int_0^\infty \exp(-sL\xi^2) \int_{\sum x_j^2 = L\xi^2} \exp\left\{\frac{\beta}{8} \sum_{i,j=1}^L \rho(i-j) x_i x_j + h \sum_{j=1}^L x_j\right\} d\sigma_{\xi\sqrt{L}}\, d\xi$$

$$= \sqrt{L} \int_0^\infty \exp(-sL\xi^2)\xi^{L-1} \int_{\sum \mu_j^2 = L} \exp\left\{\frac{\beta\xi^2}{8} \sum_{i,j=1}^L \rho(i-j) \mu_i \mu_j + h\xi \sum_{j=1}^L \mu_j\right\} d\sigma\, d\xi,$$

where we have first made the substitution

$$r = \xi\sqrt{L}$$

and then

$$x_j = \xi\mu_j.$$

Thus we have

$$\int_{-\infty}^\infty \cdots \int_{-\infty}^\infty \exp\left(-s \sum_{j=1}^L x_j^2\right) \exp\left\{\frac{\beta}{8} \sum_{i,j=1}^L \rho(i-j) x_i x_j + h \sum_{j=1}^L x_j\right\} dx_1, \ldots, dx_L$$

$$= \sqrt{L} \int_0^\infty \exp(-Ls\xi^2)\xi^{L-1} \tilde{F}(h\xi; \beta\xi^2)\, d\xi.$$

On the other hand, the multiple integral can be evaluated directly and the answer is

$$(\sqrt{\pi})^L \frac{e^{(h^2/4)(e)[sI - (\beta/8)R]^{-1}(e)}}{\sqrt{\det(sI - (\beta/8)R)}}$$

where the $L \times L$ matrix R is

$$R = \begin{pmatrix} \rho(0), \rho(1), \ldots, \rho(L-1) \\ \rho(1), \rho(2), \ldots, \rho(L-2) \\ \cdots\cdots\cdots\cdots \\ \rho(L-1), \ldots, \rho(1), \rho(0) \end{pmatrix}$$

and the vector (e) is the vector $(1, 1, \ldots, 1)$.

Combining the two results we have

$$\sqrt{L}\int_0^\infty e^{-Ls\xi^2}\xi^{L-1}\tilde{F}(h\xi, \beta\xi^2)\,d\xi = (\sqrt{\pi})^L \frac{e^{(h^2/4)(e)[sI-(\beta/8)R]^{-1}(e)}}{\sqrt{\det[sI-(\beta/8)R]}} \tag{2.2.13}$$

and it is clear that the result makes sense only if

$$s > \frac{\beta}{8} \times \text{largest eigenvalue of } R.$$

By a classic result of Szegö

$$\lim_{L\to\infty}\frac{1}{L}\det\log\left(sI - \frac{\beta}{8}R\right) = \frac{1}{2\pi}\int_{-\pi}^{\pi}\log\left(s - \frac{\beta}{8}g(\theta)\right)d\theta \tag{2.2.14}$$

where

$$g(\theta) = \sum_{-\infty}^{\infty}\rho(k)e^{ik\theta}$$

and

$$s > \frac{\beta}{8}g(0) = \frac{\beta}{8}\max g(\theta)$$

(it follows also from Szegö's result that $\lambda_{\max}(R) \to g(0)$ as $L \to \infty$).

One can also prove that

$$\lim_{L\to\infty}\frac{1}{L}(e)\left(sI - \frac{\beta}{8}R\right)^{-1}(e) = \frac{1}{s - (\beta/8)g(0)} \tag{2.2.15}$$

and finally

$$\lim_{L\to\infty} \frac{1}{L} \log \int_0^\infty e^{-sL\xi^2} \xi^{L-1} \tilde{F}(h\xi, \beta\xi^2)\, d\xi$$

$$= \frac{1}{2}\log \pi + \frac{h^2}{4}\frac{1}{s - (\beta/8)g(0)} - \frac{1}{2}\frac{1}{2\pi}\int_{-\pi}^{\pi} \log\left(s - \frac{\beta}{8} g(\theta)\right) d\theta. \tag{2.2.16}$$

One can prove that

$$\lim_{L\to\infty} \frac{1}{L} \log \tilde{F}(h\xi, \beta\xi^2) = \tilde{f}(h\xi, \beta\xi^2)$$

exists and by the usual argument

$$\operatorname*{Max}_{\xi}\{-s\xi^2 + \log \xi + \tilde{f}(h\xi, \beta\xi^2)\}$$

$$= \frac{1}{2}\log \pi + \frac{h^2}{4}\frac{1}{s - (\beta/8)g(0)} - \frac{1}{2}\frac{1}{2\pi}\int_{-\pi}^{\pi} \log\left(s - \frac{\beta}{8} g(\theta)\right) d\theta.$$

For a given s (and h and β) we find $\xi(s)$ which maximizes the expression in braces from the equation

$$-2s\xi + \frac{1}{\xi} + \frac{d}{d\xi}\tilde{f}(h\xi, \beta\xi^2) = 0 \tag{2.2.17}$$

and then we have

$$-s\xi^2(s) + \log \xi(s) + \tilde{f}(h\xi(s), \beta\xi^2(s))$$

$$= \frac{1}{2}\log \pi + \frac{h^2}{4}\frac{1}{s - (\beta/8)g(0)} - \frac{1}{2}\frac{1}{2\pi}\int_{-\pi}^{\pi} \log\left(s - \frac{\beta}{8} g(\theta)\right) d\theta. \tag{2.2.18}$$

Differentiating with respect to s and taking into account (2.2.17) we finally obtain

$$\xi^2(s) = \frac{h^2}{4}\frac{1}{[s - (\beta/8)g(0)]^2} + \frac{1}{2}\frac{1}{2\pi}\int_{-\pi}^{\pi} \frac{d\theta}{s - (\beta/8)g(\theta)}.$$

Since we are interested in $\tilde{f}(h, \beta)$, i.e., we want $\xi = 1$, we must consider the equation

$$1 = \frac{h^2}{4} \frac{1}{[s - (\beta/8)g(0)]^2} + \frac{1}{2} \frac{1}{2\pi} \int_{-\pi}^{\pi} \frac{d\theta}{s - (\beta/8)g(\theta)}, \tag{2.2.19}$$

determine s from it, and then

$$\tilde{f}(h, \beta) = \frac{1}{2} \log \pi + s + \frac{h^2}{4} \frac{1}{s - (\beta/8)g(0)}$$
$$- \frac{1}{2} \frac{1}{2\pi} \int_{-\pi}^{\pi} \log\left(s - \frac{\beta}{8} g(\theta)\right) d\theta.$$

Considering s as a function of h and differentiating with respect to h we get

$$\frac{\partial \tilde{f}}{\partial h} = \frac{ds}{dh}\left(1 - \frac{h^2}{4} \frac{1}{[s - (\beta/8)g(0)]^2} - \frac{1}{2} \frac{1}{2\pi} \int_{-\pi}^{\pi} \frac{d\theta}{s - (\beta/8)g(\theta)}\right)$$
$$+ \frac{h}{2} \frac{1}{s - (\beta/8)g(0)}$$
$$= \frac{1}{2} \frac{h}{s - (\beta/8)g(0)}.$$

It may appear that as $h \to 0$ $s(h) \to s(0)$ which is the solution of the equation

$$1 = \frac{1}{2} \frac{1}{2\pi} \int_{-\pi}^{\pi} \frac{d\theta}{s(0) - (\beta/8)g(\theta)}$$

or

$$\frac{\beta}{8} = \frac{1}{2} \frac{1}{2\pi} \int_{-\pi}^{\pi} \frac{d\theta}{\zeta - g(\theta)}, \qquad \zeta = \frac{8s(0)}{\beta} \geqslant g(0),$$

and that consequently

$$\lim_{h \to 0} \frac{\partial \tilde{f}}{\partial h} = 0.$$

This is indeed the case if

$$\frac{1}{2}\frac{1}{2\pi}\int_{-\pi}^{\pi}\frac{d\theta}{g(0)-g(\theta)}=\infty$$

for then the equation

$$\frac{\beta}{8}=\frac{1}{2}\frac{1}{2\pi}\int_{-\pi}^{\pi}\frac{d\theta}{\zeta-g(\theta)}$$

has a unique solution $\xi(\beta)$ for every $\beta > 0$ and $s(h) \to s(0) = \beta\xi(\beta)/8$ as $h \to 0$.

However, if

$$\frac{1}{2}\frac{1}{2\pi}\int_{-\pi}^{\pi}\frac{d\theta}{g(0)-g(\theta)}=\frac{\beta_c}{8}<\infty,$$

then for $\beta > \beta_c$ it can be seen from (2.2.19) that

$$\left(\frac{\partial \tilde{f}}{\partial h}\right)^2=\frac{1}{4}\frac{h^2}{[s-(\beta/8)g(0)]^2}\sim 1-\frac{\beta_c}{\beta},\qquad h\to 0$$

and consequently

$$\lim_{h\to 0+}\frac{\partial\tilde{f}}{\partial h}=\sqrt{1-\frac{\beta_c}{\beta}}.$$

Thus for $\rho(k)$ leading to the convergence of the integral

$$\int_{-\pi}^{\pi}\frac{d\theta}{g(0)-g(\theta)}$$

the spherical model exhibits a phase transition.

A sufficient condition on $\rho(k)$ is

$$\rho(k)\sim\frac{1}{k^{1+\varepsilon}},\qquad 0<\epsilon<2$$

which is also a sufficient condition found by Dyson.*

---

* That the condition $\rho(k) \sim k^{-(1+\epsilon)}$, $0 < \epsilon < 2$ is also sufficient for the phase transition of the lattice gas has been conjectured on the basis of the above result for the spherical model.

REFERENCES

1. G. E. Uhlenbeck and G. W. Ford, *Lectures in Statistical Mechanics*, American Mathematical Society, Providence, R.I., 1963.
2. M. E. Fisher, "The free energy of a macroscopic system," *Arch. Rational Mech. Anal.*, **17** (1964), 377–410.
3. H. S. Green, "A general kinetic theory of liquids, II: Equilibrium properties," *Proc. Royal Soc. London Ser. A*, **189** (1947), 103–117.
4. L. Van Hove, "Sur l'intégrale de configuration pour les systèmes de particules à une dimension," *Physica*, **16** (1950), 137–143.
5. M. Kac, "On the partition function of a one-dimensional gas," *Phys. Fluids*, **2** (1959), 8–12.
6. M. Kac, P. C. Hemmer, and G. E. Uhlenbeck, "On the van der Waals theory of vapor–liquid equilibrium. I," *J. Math. Phys.*, **4** (1963), 216–228.
7. L. S. Ornstein, Dissertation, Leiden, 1908.
8. G. E. Uhlenbeck, P. D. Hemmer, and M. Kac, "On the van der Waals theory of vapor–liquid equilibrium. II," *J. Math. Phys.*, **4** (1963), 229–247.
9. T. D. Lee and C. N. Yang, "Statistical theory of equations of state and phase transitions. II: Lattice gas and Ising model," *Phys. Rev.*, **87** (1952), 410–419.
10. F. J. Dyson, "Existence of a phase transition in a one-dimensional Ising ferromagnet," *Comm. Math. Phys.*, **12** (1969), 91–107.

A SURVEY OF SOME RECENT RESULTS IN ERGODIC THEORY*

Donald S. Ornstein

MEASURE PRESERVING TRANSFORMATIONS

A large and central part of ergodic theory deals with abstract mathematical objects called measure preserving transformations (or flows) of a measure space. By a measure space (or probability space) we mean a set of points (like the points on the unit interval) together with a collection of subsets called measurable sets (like the Lebesgue measurable subsets of the unit interval), and a measure or probability assigned to each of these subsets (like Lebesgue measure). By a measure preserving transformation on a measure space we mean a mapping which assigns to each point in the measure space another point in a one to one, onto way, and so that each measurable set is transformed onto a measurable set of the same measure. By a flow

* Research supported in part by NSF MCS 76-09159.

we mean a family T_t (t a real number) of transformations such that $T_t \cdot T_s = T_{t+s}$. ($T_t \cdot T_s$ denotes the transformation obtained by first applying T_s and then T_t.)

Origins in mechanical systems. A standard mathematical approach to a problem is to abstract the essential features of a problem and then study the resulting abstract mathematical object. The attempt to study the statistical properties of mechanical systems has led to the study of measure preserving transformations or flows.

Take the case of a billiard ball moving on a square table with a finite number of convex obstacles. If we assume that the ball always moves with unit speed, then we can describe the configuration of the system by the position of the ball and the direction in which it is moving (that is, by a point in a three-dimensional manifold called the phase space of the system). The time evolution of the system is represented by the movement of this point in phase space. (If x is the point in the phase space representing our system at time 0, then the point representing our configuration at time t will be denoted by $S_t(x)$. It is clear that $S_{t_1}(S_{t_2}(x)) = S_{t_1+t_2}(x)$.) It turns out that Lebesgue measure on the phase space is preserved by the flow (that is if E is a measurable set in phase space then the measure of E and $S_t(E)$ are the same).

The measure of a set E in phase space is supposed to model the probability that the configuration of the system lies in E. The justification for this is the following: We first assume that if the Lebesgue measure of E is 0, then the configuration will lie in E with probability 0. This seems reasonable and we take it on faith. Now because of the ergodic theorem and because of Sinai's theorem on the ergodicity of the above billiard system we get that, except for a set of points x in the phase space of Lebesgue measure 0, the fraction of time that $S_t(x)$ spends in E tends to the Lebesgue measure of E as t tends to ∞.

Another mechanical system (that we will discuss later and refer to as geodesic flow on a surface) is the motion of a particle constrained to move on a compact surface. If the particle is moving at unit speed, then the configuration can be described by a point on the surface and a direction. (The phase space is thus the unit tangent space to the surface.) If the particle is at a point moving in a certain direction, it will move along the geodesic through that point in the given direction and this will determine its position and direction t units of time later. Lebesgue measure on the phase space will be preserved.

A case of special interest will be when the surface has negative curvature. This represents a physical system, because even though compact surfaces of negative curvature cannot be imbedded in three-space, Kolmogorov pointed out the same flow can be obtained by a particle constrained to move on a closed surface in three-space under the additional influence of a finite number of centers of attraction and repulsion placed near the surface.

Another system is the motion of hard spheres moving in a rectangular box. The configuration is described by a point in a high-dimensional Euclidean space by writing down the coordinates of the position and velocity of each sphere; Lebesgue measure will be preserved.

Isomorphism. It is felt that a measure preserving flow (sometimes called an abstract dynamical system) is a good model for the statistical properties of the time evolution of a mechanical system and that if two systems give rise to the same abstract flow then they are in "some sense" statistically the same.

The idea of isomorphism is a precise way of formulating the above (or of telling us what to ignore, like events of probability 0 or topological structure, etc.). We say that S_t (or T) acting on X is isomorphic to (or statistically the same as) $\bar{S}_t$ (or $\bar{T}$) acting on $\bar{X}$, if X and $\bar{X}$ are each the union of two disjoint invariant sets X_1, X_2 and $\bar{X}_1$, $\bar{X}_2$, where X_2 and $\bar{X}_2$ have measure 0, and if there is a 1–1 invertible measure preserving map φ from X_1 onto $\bar{X}_1$ such that $\bar{S}_t(\varphi(x)) = \varphi(S_t(x))$ for all x in X_1 and all t (or $\bar{T}(\varphi(x)) = \varphi\bar{T}(x)$ for all x in X_1).

Origins in random processes. Measure preserving transformations also arise in the study of stationary random processes. A stationary random process can be thought of as a box that prints out one letter each unit of time, where the probability of printing out a given letter may depend on the letters already printed out but is independent of the time (that is, the mechanism in the box does not change).

Example 1. The box contains a roulette wheel. We spin the wheel once each unit of time, and print out the result. (We call such a process an independent process.)

Example 2. The box contains a roulette wheel. We look at all possibilities for three consecutive spins of the wheel, and divide these into two classes. Each time we spin the wheel, we look at the last three spins and print out 1 if they fall in the first class, and 2 if they fall in the second class. (We call this process a finite coding of an independent process.)

Example 3. The box contains two coins, one of which is biased so that the probability of heads is not 1/2. We divide all sequences of heads and tails of length three into two classes. At each unit of time, we look at the sequence of heads and tails which the box has printed out in the last three times. If the sequence lies in the first class, we flip the first coin and print out heads if it comes up heads, and tails if it comes up tails. If the previous three print-outs were in the second class, then we would use the second coin. This is an example of a three-step Markov process; that is, the last three print-outs determine the probability of printing out "heads," but if we know the last three print-outs, the conditional probability of "heads" is unaffected by any additional knowledge of what the process printed out in the past. (Note that Example 2 need not be an n-step Markov process for any n.)

Example 4. The box contains a mechanical system such as a gas. We make a fixed measurement on the system at each unit of time which has only a finite number of possible outcomes and print out the outcome of the measurement. (If the outcome of the measurement were real-valued, we could divide the line into a finite number of sets and print out which set the measurement fell into.)

Example 5. A teleprinter. This prints out letters where the probability of printing out a given letter depends on what has already been printed. (Many possibilities will have probability 0 because they will not make sense.)

The mathematical model for a stationary random process is a measure preserving transformation of a measure space and a partition of the measure space into a finite number of disjoint sets. We can see this as follows: If we let the process run forever (with no beginning and no end), we get an infinite sequence of outputs. These sequences will be the points of our measure space. The probabilities (or measures) of those sets of sequences, formed by specifying the print-out at some finite number of specified times, is determined by (and determines) the physical process and determines the measure on our measure space. The transformation corresponds to the progression of time and thus shifts each sequence. The partition is the partition according to the output at time 0.* If we are given only the transformation and the partition, then for each point in our measure space we can recover the sequence of outputs to which it corresponded by seeing which element of the partition the point is in at various times (that is, we apply different powers of the transformation to our point and see in what element of the partition the resulting point lies).

This point of view goes back to Kolmogorov, and is especially natural if our process is repeating a measurement on a mechanical system (with a finite number of possible outcomes). Then our measure space and transformation are already given. Our partition is the one formed by putting all those configurations for which the measurement has the same outcome into one element.

Origins in abstract mathematics. Any automorphism of a compact group is a measure preserving transformation (with respect to Haar measure).

---

* Our process can then be thought of as follows: A sequence is picked at "random" according to the above measure. At time 0 we observe the 0th coordinate. We shift the sequence to the left and observe the 0th coordinate at time one, etc.

For example, the two-dimensional torus, T_2, can be regarded as R^2 modulo Z^2. Then the matrix $\begin{pmatrix} 1 & 2 \\ 1 & 1 \end{pmatrix}$ gives us an automorphism of T_2.

Translations of compact groups such as a rotation of the circle give measure preserving transformations.

A ROUGH CLASSIFICATION OF TRANSFORMATIONS

In studying the statistical properties of a mechanical system we are led to studying the properties of stationary processes that will be shared by all the measurements we make on it.

Deterministic. We say that a process is deterministic if knowledge of all of its past outputs allows us to predict (with probability one) all of the future outputs. (A formal way of saying this is the following: If we represent our process by a transformation T and partition P,* and $\bigvee_{-\infty}^{-1} T^iP$ denotes the σ-algebra generated by the T^iP, $-\infty < i < -1$, then P will be contained in $\bigvee_{-\infty}^{-1} T^iP$.) We call a transformation, T, deterministic if for all P, P,T is deterministic.

It turns out that if the T^iP generate the entire σ-algebra and if P,T is deterministic, then T is deterministic. This implies that we can characterize the deterministic transformations as those transformations that arise from deterministic processes (that is, start with a process and form the transformation in the way indicated above).

A flow is deterministic if each of its T_t is. (Although we have not yet defined entropy, it might be worth mentioning the fact that P,T is deterministic if and only if its entropy is 0.)

The simplest example of an ergodic transformation that is deterministic is the rotation of a circle through an irrational angle. In fact, any translation of a compact group is deterministic.

***K* or completely nondeterministic.** A transformation T is said to be completely nondeterministic if there are no partitions P such that the process P,T is deterministic.

---

* All partitions will be assumed to be finite.

There is another way to characterize these transformations in terms of the 0–1 law. A process satisfies the 0–1 law if the only events that can be predicted (even partially) from knowledge of what the process did in the arbitrarily distant past have probability 0 or 1 (more formally, the intersection of the σ-algebras $\bigvee_{-\infty}^{-n} T^i P$, $1 \leqslant n < \infty$ contains only sets of measure 0 or 1). A beautiful result of Sinai and Rochlin says:

THEOREM: *T is completely nondeterministic if and only if for all partitions P, P,T satisfies the* 0–1 *law. Furthermore, if a generating partition satisfies the* 0–1 *law then all partitions do.*

We can thus characterize these transformations as those arising from processes satisfying the 0–1 law.

A flow is completely nondeterministic if each transformation in it is.

Completely nondeterministic transformations were first introduced by Kolmogorov and are now usually called K-automorphisms (or K transformations).

Many natural examples have been shown to be K. Rochlin and Yuzvinskii showed that:

THEOREM [R1] [J]: *Any ergodic automorphism of a compact group is K.*

Sinai and Anosov showed that

THEOREM [S–1]: *The geodesic flow on a surface of negative curvature is K.*

THEOREM [S–2]: *The motion of a billiard ball on a square table with convex obstacles is K.*

Sinai also showed that

THEOREM [S–2]: *The motion of hard spheres in a box is K.*

Bernoulli shifts. The most random transformations possible, the Bernoulli shifts, are those transformations that arise from independent processes (see Example 1). More specifically: Start with a "roulette wheel" with k slots, and call the fraction of the circumference occupied by the ith slot p_i. This gives us an independent process and from this process we get a measure preserving transformation. Thus, if we start with k positive numbers $p_1, \ldots, p_k$ that add up to 1, we construct a transformation called the Bernoulli shift, $(p_1, \ldots, p_k)$ or $B_{(p_1, \ldots, p_k)}$. (More formally, let Y be a space with k points having measure $p_1, \ldots, p_k$, let Y_i be copies of Y, and let $X = \prod_{-\infty}^{\infty} Y_i$ be the product space with the product measure. Then $B_{(p_1, \ldots, p_k)}$ is the shift on X.)

It is not hard to see that Bernoulli shifts have the K property. The Bernoulli shift $(p_1, \ldots, p_k)$ is isomorphic to the following transformation of the unit square. We divide the square into k rectangles where the ith rectangle has a base of width p_i and height 1. We squeeze the height of the ith rectangle and expand its width so that it now has height p_i and width 1. We now reassemble these rectangles to form the square again by putting the first on the bottom, the second on top of it, etc.

We will draw the picture for $B_{(\frac{1}{2}\ \frac{1}{2})}$.

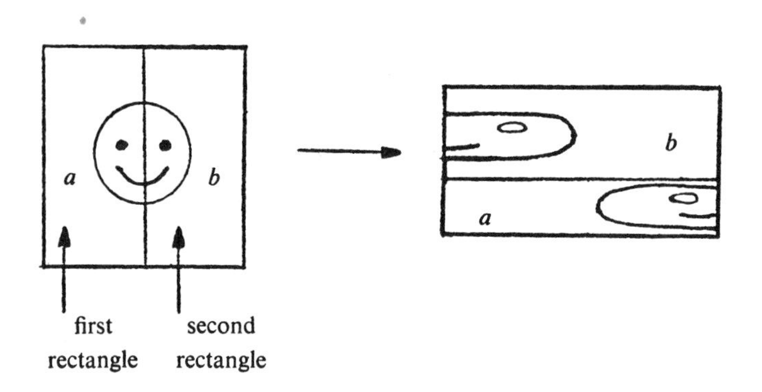

What might have been. It was once thought that the class of K automorphisms was the same as the class of Bernoulli shifts (we already noted that Bernoulli implies K). It was also thought that every transformation was either deterministic or K (completely non-deterministic) or the direct product of a deterministic transformation and a K automorphism (transformation). This was the state of affairs for some time and only relatively recently was it shown to be false.

THEOREM [O–1]: *There exists a K automorphism that is not (isomorphic to) a Bernoulli shift.*

THEOREM [O–2], [O–3]: *There is a transformation that is not deterministic or K or a direct product of a deterministic transformation and a K automorphism.*

ENTROPY AND THE NONISOMORPHISM OF BERNOULLI SHIFTS

If we were describing our subject in the order in which it actually developed we would have started with the oldest of the isomorphism problems for really random transformations: are all Bernoulli shifts isomorphic? One of the first results in ergodic theory was an isomorphism theorem for certain deterministic transformations due to Von Neuman and Halmos: If the unitary operator V on L_2 induced by T has the property that the eigenvectors span L_2, then V determines T. There were also some qualitative measures of randomness, such as mixing, weak mixing, etc., which distinguished some transformations. However, there was a large class of transformations (which included the Bernoulli shifts and K automorphisms) that possessed all of the known mixing properties and they all induced the same unitary operator. There was no way to distinguish these really random transformations and people focused attention on the simplest case, namely the isomorphism problem for Bernoulli shifts.

The breakthrough came in 1956 when Kolmogorov introduced the invariant called entropy which was motivated by Shannon's work on information theory. This invariant could be computed for the Bernoulli shift $p_1, \ldots, p_k$ and turned out to be $-\Sigma p_i \log p_i$. (Thus, the Bernoulli shift $(\frac{1}{2}\ \frac{1}{2})$ is not isomorphic to the Bernoulli shift $(\frac{1}{3}\ \frac{1}{3}\ \frac{1}{3})$.)

It was the entropy theory that motivated the definition of deterministic (entropy 0), and the K property that made the Sinai-Rochlin theorem possible and made the splitting conjecture reasonable.

The entropy of a transformation is based on the entropy of a stationary process P,T. There is a theorem due to Shannon and MacMillan which says that there is a positive real number, associated with P,T (called the entropy of P,T or $H(P,T)$), such that given ϵ there is an M, and for each $n > M$ we can divide the sequences of length n that the process prints out (i.e., the atoms in $\bigvee_1^n T^iP$) into two sets. The probability of an n-string being in the first set is less than ϵ and any string in the second set has probability between $\frac{1}{2}^{(H(P,T)\pm\epsilon)n}$. Kolmogorov and Sinai defined the entropy of T, $H(T)$, to be the sup over all finite P of $H(P,T)$. They then showed that if the T^iP generate the entire σ-algebra, then $H(P,T) = H(T)$. Because it is easy to calculate the entropy of an independent process, the above results allowed them to calculate the entropy of Bernoulli shifts.

ISOMORPHISM

The isomorphism problem can now be stated: when are two transformations of the same entropy isomorphic?

The first general isomorphism theorem was proved by Adler and Weiss:

THEOREM [A–W]: *Ergodic group automorphisms of the two-dimensional torus having the same entropy are isomorphic.*

Returning to Bernoulli shifts we observe that there are a lot of Bernoulli shifts of the same entropy, and the question remains: are they isomorphic? A lot of work was done on this problem, and in particular, Sinai proved a deep and beautiful theorem which inspired all of the later work along these lines.

THEOREM [S–3]: *Let T be any ergodic transformation and $p_1, \ldots, p_k$ positive numbers such that $-\Sigma p_i \log p_i \leqslant H(T)$. Then there is a partition P whose atoms have measure $p_1, \ldots, p_k$ and such that the T^iP are independent.*

We should note that if the T^iP generated (i.e., generated the σ-algebra), then T would be isomorphic to a Bernoulli shift.

We now have

THEOREM [O–4]: *Two Bernoulli shifts of the same entropy are isomorphic.*

The above theorem is a special case of a more general isomorphism theorem which we will describe in the next section. Before doing this, however, we will give some consequences of the general theorem.

THEOREM [O–5]: *There is a flow, S_t such that S_1 is a Bernoulli shift.* (Call such a flow a Bernoulli flow.)

One consequence of this theorem is that the Bernoulli shift $(\frac{1}{2}, \frac{1}{2})$ has a square root. This fact is also an immediate consequence of the isomorphism of Bernoulli shifts, but it was not known before this was proved.

THEOREM* [O–6]: *There is only one Bernoulli flow. That is (except for a trivial change in the time scale), any two Bernoulli flows are isomorphic.*

It follows from the above theorem and properties of a special model that if S_1 is Bernoulli then S_t is Bernoulli for all t.

One model for the Bernoulli flow (first studied by Kakutani and proved by Gurevich [Gu] to be K) has a simple description. A continuous time stationary process gives a flow just as a discrete time stationary process gives a transformation. The continuous time process we start with to get the Bernoulli flow is the following. The process puts out two signals, 0 and 1, 0 always lasts α units of time and 1 always

---

* If we consider independent processes with an infinite number of possible outcomes we get Bernoulli shifts of infinite entropy. There are then two Bernoulli flows, one of finite and one of infinite entropy. However, for the sake of the exposition we will ignore this by only considering transformations that have a finite generating partition.

lasts β units of time, and α/β is irrational. At the end of each signal we flip a coin. If the result is heads we produce a 1, if it is tails we produce a 0. (There will be no break between successive 1's (or 0's).)

A more formal description of the above flow is the following. Let T acting on X be the 2-shift. Let f be the function on X that takes on two values: α on those points of X whose first coordinate is 0 and β on those points whose first coordinate is 1. α and β are picked so that α/β is irrational. Let Y be the area under the graph of f. S_t will act on Y as follows: each point, (x, y), will move directly up at unit speed until it hits the graph of f. Then it goes to $(Tx, 0)$ and continues moving up at unit speed. S_t, for each t, is shown to be a Bernoulli shift.

Another consequence of the general isomorphism theorem is that if T is Bernoulli and α any σ algebra invariant under T (except the trivial σ algebra of sets of measure 0 or 1), then we can find a partition P in α such that the T^iP are independent and generate α. T restricted to α is called a factor of T and the above can be summed up as follows:

THEOREM [O–7]: *A factor of a Bernoulli shift is Bernoulli.*

One of the main points of the general isomorphism theorem is that it gives us a criterion that allows us to prove that some specific systems are Bernoulli (and thus determine their structure up to isomorphism). It is not always easy to verify this criterion and in many cases a very deep analysis was needed to do this. I will now list some of the systems that have been shown to be Bernoulli.

(1) *The transformations arising from mixing multi-step Markov processes on a finite statespace.* (This result was first proved by N. Friedman and myself.) [F–O]

(2) *Ergodic automorphism of a compact group.* This theorem in its full generality was proved by Thomas and Miles [T–M] and by Lind [L–1]. The proofs are different, but both involve a very deep analysis of the structure of group automorphisms. The problem has a long history and there were many partial results along the way. Halmos [H] first proved that any ergodic group automorphism was

mixing. Rochlin [R–1] and Yusvinskii [Y] proved it had the K property. Weiss and I [W–O–1] showed that the ergodic automorphisms of the 2 torus were Bernoulli and Katznelson [K] extended this to the n-torus. Lind [L–2] and Totoki [T] extended this to the infinite dimensional torus.

(3) *The motion of a billiard ball on a square table with a finite number of convex obstances and the hard sphere gas confined in a box at a fixed energy.* Sinai, by a very hard and deep analysis of these systems, was able to prove ergodicity and even the K property [S–2]. Gallavotti and I [G–O] showed that his methods give Bernoulliness for the billiard system. Sinai (oral cummunication) showed that his methods give Bernoulliness for hard spheres.

(4) *Geodesic flow on a surface of negative curvature.* This was proved to be Bernoulli by Weiss and myself [W–O–1] using results of Sinai, Anosov, and Hopf about geodesic flow. The study of this system has a long history. Hopf [Hf], Hedlund [Hd], Gel'fand and Fomin [G–F], Sinai [S–1] and Anosov [A] successively proved it to be more and more random, establishing ergodicity, mixing, Lebesgue spectrum and K.

The above four systems are by no means a complete list. There are analogous results for certain hyperbolic systems and also results about number theoretic transformations (many to one transformations of the unit interval), see 1–22.

A DISTANCE BETWEEN PROCESSES AND THE GENERAL ISOMORPHISM THEOREM

We would like to describe here the general isomorphism theorem to which we alluded in the previous section which allowed us to get our theorems for flows and from which we derived the Bernoulliness of specific examples. Before doing this we must describe a distance between processes.

There is a natural notion of distance between processes (which we will call $\bar{d}$ distance) that we will need in our description of measurements. (Roughly speaking, if we change a process very infrequently, then the new process should be close to the old one.) We say that two processes have distance less than ϵ if and only if there exists a stationary process that prints out two letters at each unit of time; and if we look only at the first letter we get the first process, and if we look only at the second letter we get the second process. Furthermore, the probability that the two letters differ is less than ϵ.

In terms of the model for a process consisting of a transformation T and an ordered partition P we have that $\bar{d}((T,P),(\bar{T},\bar{P})) < \epsilon$ if and only if there is an ergodic transformation T_1 and partitions P_1 and $\bar{P}_1$ such that the measure of the set of points labeled differently by P_1 and $\bar{P}_1$ is less than ϵ, and P_1,T_1 and $\bar{P}_1,T_1$ give the same process as P,T and $\bar{P},\bar{T}$ respectively. (That is, for each n corresponding atoms in $\bigvee_1^n T^iP$ and $\bigvee_1^n T_1^iP_1$ have the same measure. Since our partitions are ordered, it makes sense to talk about corresponding atoms.)

There is another definition in terms of typical sequences (where a typical sequence for a process is a sequence such that each finite string occurs with a frequency equal to its probability). If two processes differ by less than ϵ, then for almost every (typical) sequence of the first process we can find a typical sequence of the second that differs from it on a set of frequency less than ϵ. Furthermore, if we can find a typical string from each process, such that the two strings differ with frequency less than ϵ then the $\bar{d}$ distance is less than ϵ.

We will say that two processes P,T and $\bar{P},\bar{T}$ (where P and $\bar{P}$ have the same number of atoms and the atoms are labeled with the same set of labels) are ϵ close for time n if any string of length n gets probabilities from P,T and $\bar{P},\bar{T}$ whose ratio is between $(1 \pm \epsilon)$.

DEFINITION: A process P,T is finitely determined or F.D. if given ϵ there is a δ and K such that if $\bar{P},\bar{T}$ is any process such that: (1) P and $\bar{P}$ have the same number of atoms labeled by the same set; (2) $H(P,T) - H(\bar{P},\bar{T}) < \delta$; and (3) P,T and $\bar{P},\bar{T}$ are δ close for time K, then $\bar{d}((P,T),\bar{P},\bar{T}) < \epsilon$.

Independent processes are F.D.

GENERAL ISOMORPHISM THEOREM [O–8]–[O–5]: *If T and $\bar{T}$ have the same entropy and both have* F.D. *generators (i.e., there are partitions P and $\bar{P}$ such that T^iP and $\bar{T}^i\bar{P}$ generate and P,T and $\bar{P},\bar{T}$ are both* F.D.), *then T and $\bar{T}$ are isomorphic.*

THEOREM [O-5]-[O-8]: *If T is a Bernoulli shift and P any (finite) partition, then P,T is* F.D.

COROLLARY: *Factors of Bernoulli shifts are Bernoulli.*

There is another condition "Very Weak Bernoulli" or V.W.B., that can be shown to be equivalent to F.D. and is the condition that is verified in the concrete examples we mentioned previously.

In order to do this we need to define the $\bar{d}$ distance between two measures u and $\bar{u}$ on sequences of symbols $1, \ldots, k$, of length n. u (and $\bar{u}$) can be described by a sequence of partitions P_i, $1 \leqslant i \leqslant n$ of a measure space X. (P_i will have k atoms. X is the space of sequences and P_i is the partition according to the ith coordinate.) *Define* $(\bar{d}\{P_i\}_1^n, \{\bar{P}_i\}_1^n)$ as follows. Let φ and ψ be 1-1 measure preserving maps of X and $\bar{X}$ onto Y. Let $\bar{d}_{\varphi\psi}(\{P_i\}_1^n, \{\bar{P}_i\}_1^n) = (1/n) \times \sum_{i=1}^n |\varphi(P_i) - \psi(\bar{P}_i)|$ ($|Q - \bar{Q}|$ denotes the measure of the points labeled differently by Q and $\bar{Q}$). Define $\bar{d}(\{P_i\}_1^n, \{\bar{P}_i\}_1^n)$ as the inf over φ, ψ of $\bar{d}_{\varphi,\psi}(\{P_i\}_1^n, \{\bar{P}_i\}^n)$.

It can be shown that $\bar{d}((P,T), (\bar{P},\bar{T}))$ as previously defined is $\lim_{n\to\infty} \bar{d}(\{T^iP\}^n\{\bar{T}^i\bar{P}\}_1^n)$.

DEFINITION: P,T is V.W.B. iff given ϵ there is an n such that for all l and all atoms $A \in \bigvee_{-l}^{0} T^iP$, except for a collection of A the measure of whose union is less than ϵ, we have $\bar{d}(\{T^iP\}_1^n, \{T^iP/A\}_1^n) < \epsilon$. ($T^iP/A$ means: normalize A to have measure 1 and consider T^iP cut down to A as a partition of A.)

V.W.B. is a weak form of independence and Very Weak Bernoulli is just bad terminology.

THEOREM [W-O-2]: *P,T is* F.D. *if and only if P,T is* V.W.B. (The if part is due to me and the only if part is due to Weiss and myself.)

RANDOM PROCESSES AGAIN

We saw earlier that the transformations T that were deterministic or K could be characterized by properties that all measurements P,T had in common. We will now do the analogous thing for Bernoulli shifts, i.e., say what it means for a system to be Bernoulli in terms of the measurements on it.

Let us define a B-process to be a process (P,T) where T is a Bernoulli shift. (Because of the factor theorem it does not matter whether or not we assume the T^iP generate; thus P,T is a B-process if and only if it is F.D. or V.W.B.

THEOREM [O–8]: *Any process that can be approximated arbitrarily well in $\bar{d}$, by B-processes is a B-process.*

THEOREM [O–8]: *The B-processes are exactly those processes that can be approximated arbitrarily well in $\bar{d}$ by finite codings of a roulette wheel* (see Example 2). (One direction is obvious; the other follows from the above theorem and the factor theorem.)

THEOREM [F–O] [O–8]: *The B-processes are exactly those processes that can be approximated arbitrarily well by multi-step mixing Markov processes* (see Example 3).

It is easy to check that multi-step mixing Markov processes are V.W.B. Furthermore, any mixing process P,T is close in entropy and in finite time k to some k step mixing Markov process. If P,T is F.D., then we have $\bar{d}$ closeness.

Most of the processes one encounters in practice are B-processes. The B-processes are the most random possible. The last theorem implies that they are the only processes where wiping out the memory of the distant past will have hardly any effect (in the $\bar{d}$ sense).

Any measurement, repeated at discrete intervals of time on a geodesic flow on a surface of negative curvature is a B-process. Thus, even though the flow evolves in a completely deterministic way, as soon as we make a gross measurement (one with only a finite number of possible outcomes, e.g., look at the system on a TV screen) and discretize time (make a movie of it), the resulting process is the most random possible, and is essentially indistinguishable from a finite coding of a roulette wheel or a multi-step Markov process. This helps explain why statistical methods can play such an important role in the study of mechanical systems.

Because the B-processes are $\bar{d}$ closed and because not all processes that satisfy the 0–1 law are B (there is a K automorphism that is not Bernoulli), we have a process that satisfies the 0–1 law but cannot be approximated by codings of a roulette wheel or a process with finite memory. This means, in some rough sense, that not all processes satisfying the 0–1 law can be regarded as having as their random mechanism a roulette wheel (or a finite memory process).

HOW BAD CAN A K-AUTOMORPHISM BE?

Since we know that the class of K-automorphisms is strictly larger than the class of Bernoulli shifts, it makes sense to investigate the qualitative differences between these classes, or the properties of Bernoulli shifts that are or are not shared by K-automorphisms.

Sinai showed a long time ago that all K-automorphisms (and hence Bernoulli shifts) gave rise to the same unitary operator on L_2 of its measure space. Furthermore, all K-automorphisms are mixing of all orders. The Sinai-Rochlin theorem also gives regularity properties shared by all K-automorphisms.

Almost all of the finer properties of Bernoulli shifts, however, are not shared by K-automorphism and we will now give a list of these.

THEOREM [O–S]: *There are uncountably many nonisomorphic K-automorphisms of the same entropy* (Ornstein and Shields).

THEOREM [Sm–2]: *There is a K-flow that is not Bernoulli and there are even uncountably many nonisomorphic K-flows having the same entropy at time one* (Smorodinsky).

THEOREM [O–S]: *There is a K-automorphism not isomorphic to its inverse* (Ornstein-Shields).

THEOREM [O–2] [O–C]: *There is a K-automorphism with no square root.* (Clark extended this to no roots at all.)

THEOREM [R–1]: *There are two nonisomorphic K-automorphisms with isomorphic squares* (Rudolf).

THEOREM [P] [R–2]: *There are two nonisomorphic K-automorphisms, each of which is a factor of the other* (Polit and Rudolf).

THE ACTION OF MORE GENERAL GROUPS

The study of a single transformation and its iterates can be thought of as the study of the action of the integers Z, while the study of flows is the study of the action of the reals R. In the case of mechanical systems the action of Z and R usually represents the passage of time. Mechanical systems can, however, have other automorphisms, and this provides some motivation for the study of the action of general groups.

Weiss pointed out that for any countable group G and probabilities $p_1, \ldots, p_k (\Sigma p_i = 1)$ we have a Bernoulli $G_{p_1}, \ldots, {}_{p_k}$ action. (This is defined as follows. Let Y have k points having measure $p_1, \ldots, p_k$. Let $X = \prod_{i=G} Y_i$ where Y_i is a copy of Y, and the measure is product measure. Thus, each $x \in X$ is a sequence $\{a_i\}_{i \in G}$, $a_i \in Y$, and if $g \in G$ then $g\{a_i\} = \{b_i\}$, $b_i = a_{i \cdot g}$. This is the usual definition when $G = Z$.)

If G is a topological group of measure preserving transformations, we call G Bernoulli if every discrete countable subgroup is Bernoulli. In this case, as in the case of R, existence requires proof.

Katznelson and Weiss [K–W] began the study of more general groups by studying the theory of the n-dimensional integers, Z^n.

In studying the action of Z, certain subsets of Z, intervals of consecutive integers, $(1, 2, \ldots, n)$ played a special role. In stating the ergodic theorem, the Shannon McMillan theorem, the Kakutani-Rochlin theorem, defining entropy, etc., we needed to talk about objects like $\{T^iP\}_1^n$ or sequences of length n (atoms in $\bigvee_0^{n-1} T^iP$), etc. In Z^2, squares (the set of pairs of integers $(i, j)|i| < n$ and $|j| < n$ or $0 \leqslant i \leqslant n, 0 \leqslant j \leqslant n$) play the role of intervals (the set of integers $|i| < n$ or $0 \leqslant i \leqslant n$).

This allows us to translate most of the theorems and definitions from actions of Z to actions of Z^n. Some of the basic ergodic theory becomes more difficult, but most of the isomorphism theory goes through [O–W–3]. For example:

(1) "$\bar{d}$" and "Finitely Determined" can be defined in an analogous way.

(2) *Two* Z^n *actions with the same entropy and having* F.D. *generators are isomorphic.*

(3) This implies that *Bernoulli actions of the same entropy are isomorphic.*

(4) *Any partition of a Bernoulli* Z^n *action is* F.D. (*Hence, factors of Bernoulli actions are Bernoulli.*)

(5) There is a definition of V.W.B. such that *V.W.B.* $\Leftrightarrow$ *F.D.* (However, since the definition of V.W.B. makes a lot of use of the linear order of the integers one must be fairly careful.)

(6) The idea of K-automorphism generalizes to Z^n actions if one takes as the definition: no partitions of 0 entropy. We then have that *there are K-actions of* Z^n *that are not Bernoulli.* Burton has shown that *there are K-actions of* Z^2 *such that no transformation in* Z^2 *is a Bernoulli shift.*

Steppen pointed out that the isomorphism theorem for Bernoulli actions of Z^n (or any countable group containing Z), follows by a simple coding argument from the isomorphism theorem for Z. However, to get the general isomorphism theorem, the theorem for factors, the V.W.B. criterion, extensions to R^n, and identification of specific examples, etc., one must develop the whole theory in the context of Z^n.

Examples of Z^3 actions are Ising spin systems. A three-dimensional Ising spin system is a three-dimensional lattice of particles, each of which has two possible spins. We can describe a configuration of the system by assigning a $+1$ or -1 to each integer point in three-space. The Ising model consists of a probability distribution on these configurations.

It is usually assumed that this measure is invariant under spacial shifts. We thus have Z^3 acting on our probability space. In many cases this action can be shown to be Bernoulli [O–W–3] [G] (by establishing V.W.B.). The implications of Bernoulliness are that the configurations are very random and lack any kind of global structure, that the states vary in a $\bar{d}$ continuous way as we vary certain physical parameters and that the infinite state is fairly well determined by the entropy and the behavior of a large finite piece. This last can be regarded as an argument for the reasonableness of the infinite model.

Lind extended the isomorphism theory to actions of the n-dimensional reals, R^n.

THEOREM [L-2]: *There exists a unique Bernoulli R^n action* (and *if any of the Z^n actions induced by R^n are Bernoulli, then they all are*).

An example of an R^3 action is a Poisson distribution of spheres in three-space. R^3 acts as translations, and the action can be shown to be Bernoulli (of infinite entropy) [L–2]. If the spheres are moving, then the system has a time evolution and we get an action of R^4. It is not known whether or not this is Bernoulli.

There are more refined results about Bernoulli shifts, some of which we will come to later, and these have not yet been investigated for actions of Z^n, R or R^n.

The class of groups for which there is an isomorphism theory is not yet known. It is known that the basic theory works for actions of a countable solvable group (Ornstein-Weiss [O–W–3], Kieffer [Ki], Krieger [Kr], although the definition of V.W.B. does not seem to generalize in a reasonable way. Also some of the theory is known to break down for actions of free groups (Rudolf, Kean, Weiss). Very little is known about actions of general topological groups.

FACTORS

By a factor we mean the action of T on an invariant sub-sigma algebra.

Twenty years ago it was not even clear that every Bernoulli shift had a proper factor. The entropy theory, of course, shows that any transformation of positive entropy has lots of factors. There is an example of a mixing transformation with no factors.

What transformations can arise as factors of a given transformation? The K-automorphisms are exactly those transformations having no factors of 0 entropy. Transformations with continuous spectrum are those having no factors isomorphic to a rotation of the circle since an eigenfunction is the same thing as a rotation factor. As we have already mentioned: any factor of a Bernoulli shift is Bernoulli.

The next kind of question and the one that we will be mainly concerned with here is: *How is a factor imbedded in a transformation?* We will say that two factors of T are imbedded in the same way if there is an automorphism of T taking one factor onto the other. We will be especially interested in the case when T is Bernoulli.

There are several qualitative ways in which a factor $\mathscr{A}$ can be imbedded.

(1) *$\mathscr{A}$ can split off* (that is, T is the direct product of $\mathscr{A}$ and an orthogonal factor $\overline{\mathscr{A}}$). If T is Bernoulli $\overline{\mathscr{A}}$ is automatically Bernoulli. Otherwise we can study the case where $\overline{\mathscr{A}}$ is Bernoulli.

(2) *Any factor properly containing $\mathscr{A}$ has strictly larger entropy* (we will say in this case that $\mathscr{A}$ is maximal given its entropy or simply maximal).

(3) *$\mathscr{A}$ has the same entropy as T.*

We will now discuss these cases in more detail.

(1) *$\mathscr{A}$ splits off with a Bernoulli complement.* A very interesting and deep recent development is Thouvenot's relativized isomorphism theory which deals with this case. If P and H are partitions, Thouvenot introduces the idea of P being "finitely determined relative to H" or "F.D. rel H." (P is F.D. rel H if given ϵ there is a δ and an n such that if $\overline{T},\overline{P},\overline{H}$ satisfies

(1) $\overline{T},\overline{H}$ and T,H represent the same process;

(2) $|H(P \vee H,T) - H(\overline{P} \vee \overline{H},\overline{T})| < \delta$; and

(3) $\bar{d}(\{T^i(P \vee H)\}_1^n, \{T^i(\overline{P} \vee \overline{H})\}_1^n) < \delta$;

then $\bar{d}_{H,\overline{H}}(P \vee H,T, \overline{P} \vee \overline{H},\overline{T}) < \epsilon$, where $\bar{d}_{H,\overline{H}}$ means that when superimposing the $T^i(P \vee H)$ on the $\overline{T}^i(\overline{P} \vee \overline{H})$ the T^iH and $\overline{T}^i\overline{H}$ must fit exactly.)

If the factor generated by P is Bernoulli and orthogonal to the factor generated by H, then P is F.D. rel H.

RELATIVIZED ISOMORPHISM THEOREM (Thouvenot) [T–1]: *If P is* F.D. rel H, $\overline{P}$ *is* F.D. rel $\overline{H}$, *$H(P \vee H,T) = H(\overline{P} \vee \overline{H},\overline{T})$, and H,T and $\overline{H}, \overline{T}$ represent the same process, then there is an isomorphism between T acting on $\bigvee_{-\infty}^{\infty} T^i(P \vee H)$ and $\overline{T}$ acting on $\bigvee_{-\infty}^{\infty} \overline{T}^i(\overline{P} \vee \overline{H})$ taking H onto $\overline{H}$.*

COROLLARY: *If $P \vee H$ generate and P is* F.D. rel H, *then $\bigvee_{-\infty}^{\infty} T^iP$ splits off with a Bernoulli complement.*

Relativizing the characterizations of partitions of Bernoulli shifts, Thouvenot gets

THEOREM [T1]: *If* $\bigvee_{-\infty}^{\infty} T^i H$ *splits off with a Bernoulli complement, then any* P *is* F.D. rel H.

The idea of V.W.B. can also be relativized [T1] [Ra] (Thouvenot-Rahe).

Here are some applications:

THEOREM (Thouvenot) [T1]: *Any factor of the direct product of a transformation* 0 *entropy and a Bernoulli shift also has this form.*

THEOREM (Thouvenot and Shields) [T–S]: *The processes arising from transformations of the above form are* $\bar{d}$ *closed.*

In the case where T is a Bernoulli shift we have

THEOREM (Thouvenot) [T1]: *If* $T^i(P \vee Q)$ *are independent, then* $\bigvee_{-\infty}^{\infty} T^i P$ *splits off.* (Rahe showed that if we lump together some states of a Markov process with no transitions of 0 probability, then the resulting factor splits off.)

THEOREM (Rahe): *If* $\mathscr{A}$ *splits off with a Bernoulli complement under* T^2, *then it does so under* T.

THEOREM: *The special examples of K-automorphism that are not Bernoulli, constructed by Shields and myself all have factors of arbitrarily small entropy that split off with a Bernoulli complement (this is called the "Weak Pinsker" property).*

Thouvenot has shown (by a very deep argument):

THEOREM [T–2]: *The weak Pinsker property is stable under the taking of factors and* $\bar{d}$ *limits.*

The main open problem in this area is whether or not every transformation of positive entropy or every K-automorphism has the weak Pinsker property.

The above gives one the feeling that if $\mathscr{A}$ splits off with a Bernoulli complement, then the way in which $\mathscr{A}$ is imbedded is in some sense Bernoulli, or that in some sense the "transformation relative to $\mathscr{A}$" is Bernoulli.

(2) *$\mathscr{A}$ is maximal given its entropy.* We will now only consider the case where T is Bernoulli. It is not hard to see that if $\mathscr{A}$ splits off, then $\mathscr{A}$ is maximal given its entropy, and it is natural to ask whether the converse is true.

THEOREM [O–9]: *There is a factor of a Bernoulli shift that is maximal given its entropy but does not split off.*

The case where $\mathscr{A}$ is maximal corresponds, in some sense, to the theory of K-automorphisms. The reasons for believing this are the following. Since $\mathscr{A}$ is not contained in a proper factor of the same entropy, the action of T relative to $\mathscr{A}$ is somehow analogous to the action of a transformation with no factors of 0 entropy. Furthermore, the factor in the above theorem is obtained by taking a skew product with a K-automorphism that is not Bernoulli and thus can in some sense be regarded as a "relativized" version of the existence of a K-automorphism that is not Bernoulli. This theorem intertwines the positive theory of Bernoulli shifts and the negative theory of K-automorphisms, and in fact we use the positive theory to prove the skew product to be Bernoulli.

The example of a transformation whose 0-entropy factor does not split off can also be thought of as a "relativized" version of a K-automorphism that is not Bernoulli.

The above analogies lead us to conjecture that there are uncountably many different ways that a maximal factor of given entropy can be embedded in a Bernoulli shift.

(3) *$\mathscr{A}$ has the same entropy as T.* In this case the "action of T relative to $\mathscr{A}$" is analogous to a 0 entropy transformation.

The most interesting case is where $\mathscr{A}$ has finite fibers. We say that $\mathscr{A}$ has n-point fibers if the points in our space can be partitioned into groups of n-points such that each set in $\mathscr{A}$ is made up of complete groups and any two groups can be separated by a set in $\mathscr{A}$. One of the deepest results in the isomorphism theory is:

THEOREM (Rudolph) [R–3]: *There are only a finite number of different ways a factor with n-point fibers can be imbedded.*

Rudolph can also give an algorithm for counting the factors. Thus, there is only one factor with two-point fibers, several factors with three-point fibers.

It may be of some interest to point out that one step in Rudolph's proof involves the isomorphism theory for groups other than Z.

EQUIVALENCE THEORY

Structure of orbits. One of the most interesting recent developments is a theory that parallels the isomorphism theory and is designed to study the orbit structure of flows.

We will say that two flows S_t and $\bar{S}_t$ have the same orbit structure if there is a 1–1 invertible measurable map φ (one that takes measurable sets to measurable sets and sets of measure 0 to sets of measure 0 but does not necessarily preserve measure) and φ maps orbits of S_t, in an order preserving way onto orbits of $\bar{S}_t$.

Another way to say the same thing would be to say the S_t and $\bar{S}_t$ are equivalent if we can get $\bar{S}_t$ from S_t by changing the speed (but not the direction) at which a point moves along the orbits of S_t (the speed change will in general vary from point to point).

The first question one asks is how many equivalence classes of ergodic flows are there (if S_t is ergodic and $\bar{S}_t$ is equivalent to S_t, then $\bar{S}_t$ is ergodic)? Until the invention of entropy theory it was thought that there was only one equivalence class (except for the trivial case of only one orbit). Entropy theory gives at least three classes, flows of 0 entropy, finite entropy and infinite entropy (a flow in one of these classes cannot be equivalent to a flow in another class).

The theory remained stuck at this point for many years until Jacob Feldman showed:

THEOREM [F]: *There are nonequivalent flows of* 0, *positive and infinite entropy.*

Rudolph later extended Feldman's example to show that:

THEOREM [R–4]: *There are uncountably many nonequivalent flows in each entropy class.*

THEOREM [R–4]: *There exists a flow that is not equivalent to the flow obtained by simply reversing its direction.*

Feldman's idea was to introduce a new metric, which we will call $\bar{f}$ and to then define V.W.B. using $\bar{f}$ instead of $\bar{d}$. He then showed that this property was stable under equivalence and he was also able to construct examples without (and with) this property. Ornstein and Weiss [O–W–4] later showed that by substituting $\bar{f}$ for $\bar{d}$ one could develop a theory of equivalence that parallels the isomorphism theory. One gets an analogue to F.D. (called F.F.), the general isomorphism theorem (two flows with F.F. generators are equivalent), etc.

Before stating these results in more detail, we will give a few consequences of the theory.

The theory singles out a class of flows called "Loosely Bernoulli" or L.B. It can be shown that this is the same as F.F. *The L.B. flows of* 0 *entropy are those equivalent to the Kronecker flow on the two-dimensional torus*: represent the torus as the plane modulo the points with integer coordinates, let V be the unit vector in the direction 1, $\sqrt{2}$, and then S_t will move each point t units in the direction of V.

One consequence of the equivalence theory is that *if we substitute any irrational number for* $\sqrt{2}$, *we get an equivalent flow.*

The L.B. flows of finite entropy are those flows that are equivalent to the Bernoulli flow (of finite entropy). The L.B. *flow of infinite entropy are those equivalent to the Bernoulli flow of infinite entropy.*

Another consequence of the equivalence theory is the following:

THEOREM: (1) *If* S_t *is equivalent to the Kronecker flow, then so are its factors.* (2) *If* S_t *is equivalent to the Bernoulli flow of finite entropy, then its factors are equivalent either to the Bernoulli flow of finite entropy or the Kronecker flow.* (3) *If* S_t *is equivalent to the Bernoulli flow of infinite entropy, then its factors are equivalent to the Kronecker flow or the Bernoulli flow of finite or infinite entropy.*

We should point out that there is no relation in general between S_t and its factors as can be seen by taking direct products of non-equivalent flows.

Another consequence is that the *direct product of a* L.B. *flow and a Bernoulli flow is* L.B. (Thus, the direct product of the Kronecker flow, the geodesic flow on a surface of negative curvature has the same orbit structure as the geodesic flow.)

It should be pointed out that the taking of direct products does not respect equivalence. In fact, there *exists a* L.B. *flow whose direct product with itself is not* L.B.

Cross-sections of flows. Poincaré pointed out that the study of flows could be partially reduced to the study of transformations in the following way: Suppose S_t is a differentiable flow on a manifold X of dimension n. We can then find a piece of surface A of dimension $n - 1$ such that every orbit hits A infinitely often and only a finite number of times in any finite time. This gives us a transformation on A: if x is in A, $T(x)$ will be the point in A which is the first return of x to A. There will be a measure on A, equivalent to surface measure and invariant under T. We will refer to T as a (smooth) cross-section of S_t. T does not determine S_t, but it does determine it up to equivalence.

We will now define cross-sections more generally and in particular without any reference to differential structure. We will start with a scheme for constructing a flow from a measure preserving transformation T. Let T act on Y. Pick an integrable positive function f on Y and let X be the subset of $Y \times R$ consisting of those points lying below the graph of f (the (y, r) where $0 \leqslant r \leqslant f(y)$). S_t will move y, r up ($S_t(y, r) = y, r + t$) at unit speed until it reaches $y, f(y)$, which will be identified with $(T(y), 0)$. The point will then continue to move up at unit speed, etc. The measure on X which is derived from the product measure on $Y \times R$ will be preserved. *We will say that T is a cross-section of a flow S_t if S_t can be represented in the above form.* (Our initial differential description describes some but not all cross-sections of a differentiable flow.) There is a theorem of Ambrose and Kakutani that says that any measurable flow has a cross-section.

If we start with a transformation T and vary the f we get exactly one equivalence class of flows. Furthermore, if S_t and $\bar{S}_t$ are equivalent, then they have exactly the same cross-sections. We thus get an equivalence relation on transformations, called Kakutani equivalence: T and $\bar{T}$ are equivalent if they are cross-sections of the same (or equivalent) flow.

The equivalence theory for flows translates into a theory of Kakutani equivalence for transformations.

$\bar{f}$, F.F., etc. We will now try to describe Feldman $\bar{f}$ and the analogies of F.D., V.W.B. and the general isomorphism theorem in terms of transformations and Kakutani equivalence.

We will now define $\bar{f}$. Let $a_1, \ldots, a_n$ and $\bar{a}_1, \ldots, \bar{a}_n$ be two sequences of length n. Define $\bar{f}(a_1, \ldots, a_n, \bar{a}_1, \ldots, \bar{a}_n)$ as follows: Let m be the largest integer such that there is a sequence $b_1, \ldots, b_m$ which is a subsequence of both $a_1, \ldots, a_n$ and $\bar{a}_1, \ldots, \bar{a}_n$ (i.e., $b_i = a_{n_i}$ and if $i > j$, then $n_i > n_j$). Let $\bar{f}(a_1, \ldots, a_n, \bar{a}_1, \ldots, \bar{a}_n) = m/n$. Now suppose we have two measures u and $\bar{u}$ on sequences of length n. We can represent u (and $\bar{u}$) by a measure space X (and $\bar{X}$) of total measure 1 and function h (and $\bar{h}$) from X (and $\bar{X}$) to sequences of length n. Now let φ be a measure preserving map of X onto $\bar{X}$. Let $\bar{f}_\varphi(u, \bar{u}) = \int_X \bar{f}(h(x), \bar{h}(\varphi(x))$. Let $\bar{f}(u, \bar{u}) = \sup_\varphi \bar{f}_\varphi(u, \bar{u})$. We define $\bar{f}$ between two processes as the limit of the $\bar{f}$ distance between the measures they induce on sequences of length n.

We can now define "finitely fixed," or F.F., which is obtained from F.D. by substituting $\bar{f}$ for $\bar{d}$. P,T is F.F. if given ϵ there is a δ and K such that if $\bar{P},\bar{T}$ satisfies:

(1) P and $\bar{P}$ have the same number of atoms;

(2) $|H(P,T) - H(\bar{P},\bar{T})| < \delta$; and

(3) $\bar{f}(\{T^iP\}_1^K, \{\bar{T}^i\bar{P}\}_1^K) < \delta$,

then $\bar{f}(P,T), (\bar{P},\bar{T}) < \epsilon$.

It is now possible to have F.F. processes of 0, finite and (if we allow countable partition) infinite entropy.

GENERAL EQUIVALENCE THEOREM [O–W–4]: *If T and $\overline{T}$ have* 0 *entropy, or positive entropy, (or infinite entropy) and both have* F.F. *generators, then they are Kakutani equivalent.*

COROLLARY [O–W–4]: *Let T have a* F.F. *generator. If $H(T) = 0$, then T is Kakutani equivalent to an irrational rotation of the circle. If $0 < H(T) < \infty$, then T is Kakutani equivalent to a Bernoulli shift (if $H(T) = \infty$ then T is Kakutani equivalent to the Bernoulli shift of infinite entropy).*

FACTOR THEOREM [O–W–4]: *If T has a* F.F. *generator, then P,T is* F.F. *for all partitions P.*

We can also define L.B., the analogy of V.W.B. (which was first defined by Feldman). We say that P,T is L.B. if given ϵ there is an n such that for all l and all but ϵ of the atoms A in $\bigvee_{-l}^{0} T^iP$ we have $\bar{f}(\{T^iP/A\}_1^m, \{T^iP\}_1^m) < \epsilon$.

THEOREM [O–W–4]: *P,T is* F.F. *if and only if P,T is* L.B.

THEOREM [O–W–4]: *The* F.F. *processes of finite entropy are exactly the $\bar{f}$ closure of the multi-step Markov processes.*

The L.B. transformations of 0 entropy are the cross-sections of the Kronecker flow. Some examples are: rotation of the circle by an irrational angle, any ergodic translation of a compact group, ergodic interval exchange transformations (Weiss) (partition the unit interval into subintervals and rearrange these), any factor of the above.

The L.B. transformations of finite entropy are those transformations that arise as cross-sections of the Bernoulli flow. Examples are: Bernoulli shifts, the direct product of a Bernoulli shift and a L.B. of 0 entropy, the K but not Bernoulli transformation constructed by Ornstein and Shields. Any factor of the above is L.B. of either finite or 0 entropy.

Kakutani equivalence and inducing. When Kakutani introduced his equivalence relation, he showed how to describe it in terms of another notion "inducing." Let T be a transformation acting on X

and let E be a subset of X. There is a measure preserving transformation T_E, mapping E onto E defined as follows: $T_E(x) = T^i(x)$, where i is the smallest integer such that $T^i(x)$ is in E (we only consider x in E). We refer to T_E as the transformation induced by T on E. Kakutani showed that T and $\bar{T}$ are Kakutani equivalent if and only if there are sets E and $\bar{E}$ and T_E is isomorphic to $\bar{T}_{\bar{E}}$.

Because of this and a simple extra argument, the equivalence theorem implies that: (1) *any irrational rotation can be induced from any other irrational rotation.* (2) *Any* L.B. *of positive entropy induces any* L.B. *of larger entropy* (but cannot possibly induce an L.B. of the same entropy, thus neither E or $\bar{E}$ can be taken to be the whole space).

Other metrics. A very interesting question is whether or not there are any other metrics besides $\bar{f}$ and $\bar{d}$ that will give some sort of theory. Thouvenot's relative theory can be viewed as arising from a certain metric on pairs of processes and this adds to the hope that there are still others.

REFERENCES

A D. V. Anosov, "Geodesic flows on closed Riemannian manifolds with negative curvature," *Proc. Steklov Inst. Math.*, **90** (1967).

A–W R. Adler and B. Weiss, "Similarity of automorphisms of the torus," *Mem. Amer. Math. Soc.*, **98** (1970).

AZ M. Aizenman, S. Goldstein, and J. L. Lebowitz, "Ergodic properties of a one-dimensional system of hard rods with an infinite number of degrees of freedom," *Comm. Math. Phys.*, **39** (1975), 389.

C J. Clark Thesis, "A Kolmogorov shift with no roots".

F J. Feldman, "New *K*-automorphisms and a problem of Kakutani," to appear in *Israel J. Math.*

G–F I. M. Gel'fand and S. V. Fomin, "Geodesic flows on manifolds of constant negative curvature," *Uspehi Mat. Nauk.*, **49** (1) (1952), 118–137.

F–O N. A. Friedman and D. S. Ornstein, "On isomorphism of weak Bernoulli transformations," *Advances in Math.*, **5** (1970), 365–394.

G G. Gallavotti, "Ising model and Bernoulli schemes in one dimension," preprint.

G. Gallavotti, F. de Liberto, and L. Russo, "Markov processes, Bernoulli schemes and Ising model," to appear in *Comm. Math. Phys.*

R. Esposito and G. Gallavotti, "Approximate symmetries and their spontaneous breakdown," to appear in *Ann. Inst. H. Poincaré.*

G. Gallavotti and D. Ornstein, "Billiards and Bernoulli schemes," *Comm. Math. Phys.*, **38** (1974), 83–101.

G–O Gallavotti and Ornstein, "The billiard flow with a convex scatterer is Bernoullian," *Comm. Math. Phys.*, **38** (1974), 83. "Lectures on the billiard," by G. Gallavotti, *Lecture Notes in Physics, Dynamical Systems, Theory and Applications*, **38**, Springer-Verlag, 236–273.

G–L–A S. Goldstein, J. L. Lebowitz, and M. Aizenman, "Ergodic properties of infinite systems," *Lecture Notes in Physics, Dynamical Systems, Theory and Applications*, **38**, Springer-Verlag, 112–143.

Gu B. M. Gurevich, "A class of special automorphisms and special flows," *Soviet Math. Doklady*, **4** (1963), 1738–1741.

———, "Some conditions for K decompositions for special flow," *Trans. of the Moscow Math. Society*, **17** (1967), 99–128.

H P. R. Halmos, "Lectures on ergodic theory," *Publ. Math. Soc. Japan*, **3**, Math. Soc. Japan, Tokyo, 1956; reprint, 1960, 54–55, MR 20 3958.

Hd G. A. Hedlund, "The dynamics of geodesic flows," *Bull. Amer. Math. Soc.*, **45** (1939), 241–260.

Hf E. Hopf, "Statistik der Geodatischen Linien in Mannigfaltigkeiten negativer Krummung," *Berichte Sachs. Akad. Wiss. Leipzig*, **91** (1939), 261–304.

I–M–T S. Ito, H. Murata, and H. Totoki, "Remarks on the isomorphism theorems for weak Bernoulli transformations in general case," *Publ. Res. Inst. Math. Sciences*, Kyoto University, **7** (3) 1972.

J S. A. Juzvinskii, "Metric properties of endomorphisms of compact groups," *Izv. Akad. Nauk SSSR Ser. Mat.*, **29** (1965), 1295–1328; English transl., *Amer. Math. Soc. Transl.*, (2) **66** (1968), 63–98, MR 33, 2798.

K Y. Katznelson, "Ergodic automorphism of T^n are Bernoulli shifts," *Israel J. Math.*, **10** (1971), 186–195.

K–W Y. Katznelson and B. Weiss, "Commuting measure preserving transformations," *Israel J. Math.*, **12** (1962), 161.

Ki J. C. Kieffer, "The isomorphism theorem for generalized Bernoulli schemes," to appear in *Advances of Math.* "A generalized Shannon-McMillan theorem for the action and amenable group on a probability space," to appear in *Ann. of Math.*

Kr W. Krieger, "On the entropy of groups of measure preserving transformations," to appear.

L O. E. Lanford, III, and J. L. Lebowitz, "Time evolution and ergodic properties of harmonic systems," *Lecture Notes in Physics, Dynamical Systems, Theory and Applications*, 38, Springer-Verlag, 144–177.

L–1 D. A. Lind, "The structure of skew products with ergodic group automorphisms," *The Inst. for Adv. Studies*, The Hebrew University of Jerusalem.

L–2 ———, "Ergodic automorphisms of the infinite torus are Bernoulli."

L–3 ———, "Isomorphism theorem for R^n," to appear.

O–1 D. S. Ornstein, "An example of a Kolmogorov automorphism that is not a Bernoulli shift," *Advances in Math.*, **10** (1973), 49–62.

O–2 ———, "A K-automorphism with no square root and Pinsker's conjecture," *Advances in Math.*, **10** (1973), 89–102.

O–3 ———, "A mixing transformation for which Pinsker's conjecture fails," *Advances in Math.*, **10** (1973), 103–123.

O–4 ———, "Two Bernoulli shifts with infinite entropy are isomorphic," *Advances in Math.*, **5** (1970), 339–348.

O–5 ———, "Imbedding Bernoulli shifts in flows," Contributions to Ergodic Theory and Probability, *Lecture Notes in Mathematics*, Springer-Verlag, 1970, 178–218.

O–6 ———, "The isomorphism theorem for Bernoulli flows," *Advances in Math.*, **10** (1973), 124–142.

O–7 ———, "Factors of Bernoulli shifts are Bernoulli shifts," *Advances in Math.*, **5** (1970), 349–364.

O–8 ———, *Ergodic Theory, Randomness, and Dynamical Systems*, Yale University Press, 1974.

O–16 P. Shields and D. S. Ornstein, "An uncountable family of K-automorphisms," *Advances in Math.*, to appear.

O–W–3 D. Ornstein and B. Weiss, "Bernoulli atoms of countable groups," to appear.

O–W–4 ———, "Equivalence of measure preserving transformation," to appear.

P S. Polit, Thesis: "Weakly Isomorphic Transformations Need Not be Isomorphic."

R–1 D. Rudolph, Thesis: "Non-Bernoulli Behavior of the Roots of K-automorphisms," "Two Nonisomorphic K-automorphisms with isomorphic squares," *Israel J. Math.*

R–2 ———, "A Technique for Counterexample Construction in Ergodic Theory," in preparation.

R–3 ———, "If a finite extension of a Bernoulli shift has no finite rotation factor, it is," submitted to *Israel J. Math.*

R–4 ———, "Non-equivalence of measure preserving transformation," to appear.

Ra M. Rahe, Thesis: "Relatively Finitely Determined Implies Relatively Very Weak Bernoulli."

R–1 V. A. Rochlin, "Metric properties of endomorphisms of compact commutative groups," *Izv. Akad. Nauk SSSR Ser. Mat.*, **28** (1964), 867–874.

R–W S. M. Rudolfer and K. M. Wilkinson, "A Number-theoretic Class of Weak Bernoulli Transformations," to appear.

S–1 Ya. G. Sinai, "Geodesic flows on compact surfaces of negative curvature," *Dokl. Akad. Nauk SSSR*, **136** (3) (1961), 549–552.

S–2 Ja. G. Sinai, "On the foundations of the ergodic hypothesis for a dynamical system of statistical mechanics," *Dokl. Nauk SSSR*, **153** (1963), 1261–1264, *Soviet Math. Dokl.*, **4** (1963), 1818–1822, MR 35 #5576.

———, "Dynamical systems with elastic reflections," *Uspehi Mat. Nauk.*, **27** (1962), 137f.

S–3 ———, "A weak isomorphism of transformations with an invariant measure," *Dokl. Akad. Nauk SSSR*, **147** (1962), 797–800, *Soviet Math. Dokl.*, **3** (1962), 1725–1729, MR 28 #516a: 28 #1247.

Sm–1 M. Smorodinsky, "β-Automorphisms are Bernoulli shifts," *Acta Math. Acad. Sci. Hungar. Tomus*, **24** (1973), 3–4.

Sm–2 ———, "Construction of K-flows," to appear in *Israel J. Math.*

T1 J.-P. Thouvenot, "Quelques propriétés des systèmes dynamiques qui se décomposent en un produit de deux systèmes dont l'un est schéma de Bernoulli," *Israel J. Math.*, **21** (1965), 177–207.

T–1 ———, "Une classe de systèmes pour lesquels la conjecture de Pinsker est vraie," *Israel J. Math.*, **21** (1965), 208–214.

T–2 ———, "Weak Pinsker property," to appear.

T–S J.-P. Thouvenot and P. Shields, "Entropy zero × Bernoulli processes are closed in the $\bar{d}$-metric."

T–M Thomas and Miles, to appear in *Advances in Math.*

T N. Aoki and H. Totoki, "Ergodic automorphisms of T^∞ are Bernoulli transformations," *Publ. of RIMS Kyoto University*, **10** (2) 1976.

W–O–1 B. Weiss and D. S. Ornstein, "Geodesic flows are Bernoullian," *Israel J. Math.*, **14** (2) (1973), 184–198.

W–O–2 ———, "Finitely determined implies very weak Bernoulli," *Israel J. Math.*, **17** (1) (1974), 94–104.

1–22 L. A. Bunimovich, "Imbedding of Bernoulli shifts in certain special flows," (in Russian), *Uspehi Mat. Nauk*, (28) **3** (1973), 171–172.

1–22 M. Ratner, "Anosov flows with Gibbs measures are also Bernoullian," *Israel J. Math.*, **17** (4) (1974), 380–391.

1–22 R. Bowen and D. Ruelle, "The ergodic theory of axiom *A* flows," *Invent. Math.*, 195.

1–22 R. Bowen, "Bernoulli equilibrium states for axiom *A* diffeomorphisms," *Math. Systems Theory*, **8** (1975), 289–294.

1–22 D. Ruelle, "A measure associated with axiom *A* attractors," *Amer. J. Math.*, 1975.

1–22 I. Kubo, "Perturbed billiard systems I. The ergodicity of the motion of a particle in a compound central field," preprint.

1–22 I. Kubo and H. Murata, "Perturbed billiard systems II. The Bernoulli property."

1–22 J. Lebowitz, S. Goldstein, and M. Aizenman, "Ergodic properties of infinite systems." Battelle Rencontres, summer 1974, to appear.

1–22 S. Goldstein, "Space-time ergodic properties of systems of infinitely many independent particles," Battelle Rencontres, summer 1974.

1–22 O. Lanford, III and J. Lebowitz, "Time evolution and ergodic properties of harmonic systems," Battelle Rencontres, summer 1974.

1–22 M. Aizenman, S. Goldstein, and J. Lebowitz, "Ergodic properties of a one-dimensional system of hard rods with an infinite number of degrees of freedom," *Comm. Math. Phys.*, **39** (1975), 289.

1–22 S. Goldstein and J. Lebowitz, "Ergodic properties of an infinite system of particles moving independently in a periodic field," *Comm. Math. Phys.*, **37** (1974), 1–18.

1–22 G. Caldiera and E. Presutti, "Gibbs processes and generalized Bernoulli flows for hard-core one-dimensional systems," *Comm. Math. Phys.*, **35** (1974), 279–286, Springer-Verlag, 1974.

1–22 M. Smorodinsky, "β-automorphisms are Bernoulli shifts," *Acta Math. Acad. Sci. Hungar. Tomus*, **24** (1973), 3–4.

1–22 K. M. Wilkinson, "Ergodic properties of certain linear mod one transformations," *Advances in Math.*, **14** (1974), 64–72.

1–22 S. M. Rudolfer and K. M. Wilkinson, "A number-theoretic class of weak Bernoulli transformations," *Math. Systems Theory*, (7) **1** (1973), Springer-Verlag, New York.

1–22 S. Ito, H. Murata, and H. Totoki, "A remark on the isomorphism theorem," "On the isomorphism theorem for weak Bernoulli transformations in general case," to appear.

1–22 K. M. Wilkinson, "Ergodic properties of a class of piecewise linear trans.", *Z. Wahrscheinlichkeitstheorie und Verw. Gebiete*, (1975), 303–328, 31–34.

1–22 R. Adler and B. Weiss, "Notes on β transformations," to appear.

1–22 R. Adler, Paper given at the Conference on Recent Advances in Topological Dynamics, Springer-Verlag, New York, 318.

1–22 G. Gallavotti, "Ising model and Bernoulli schemes in one dimension," *Comm. Math. Phys.*, **23** (2) (1973), 183–190.

1–22 G. Gallavotti, F. de Liberto, and L. Russo, "Markov processes, Bernoulli schemes and Ising model," *Comm. Math. Phys.*

AUTHOR INDEX

SUBJECT INDEX